W9-CGQ-267

HUMAN CAPACITIES AND MORAL STATUS

Philosophy and Medicine

VOLUME 108

For further volumes:
http://www.springer.com/series/6414

HUMAN CAPACITIES AND MORAL STATUS

by

RUSSELL DISILVESTRO
California State University, Sacramento, CA, USA

Russell DiSilvestro
Department of Philosophy
California State University
Sacramento
6000 J Street
Sacramento CA 95819
Mendocino Hall 3016
USA
rdisilv@csus.edu

ISSN 0376-7418
ISBN 978-90-481-8536-8 e-ISBN 978-90-481-8537-5
DOI 10.1007/978-90-481-8537-5
Springer Dordrecht Heidelberg London New York

Library of Congress Control Number: 2010920978

Printed on acid-free paper

Springer is part of Springer Science+Business Media (www.springer.com)

For Gabriel
Not here but not forgotten

Acknowledgments

I would like to thank several groups of people, without whom the writing of this book would have been much more difficult, if not impossible.

First, three of my philosophy professors from my time as an M. A. student at Biola University: J. P. Moreland and Scott B. Rae, whose 2000 book *Body and Soul* persuaded and motivated me to develop the main ideas of this book, and Garry DeWeese, whose friendship and advice the last 10 years have been instrumental in bringing these ideas to print.

Second, a number of people helped an earlier draft of this book take shape as my doctoral dissertation at Bowling Green State University. My dissertation committee at Bowling Green State University generously gave their time and effort in various ways to help the dissertation come together: R. G. Frey, Fred Miller, Loren Lomasky, Michael Tooley, and Younghee Kim. The members of Fred Miller's dissertation reading group gave me helpful feedback on each part of the dissertation over a period of several months: Pete Cellelo, Kathy Erbeznik, Leanne Kent, Nico Maloberti, Chris Metivier, John Milliken, Jonathan Miller, Sangeeta Sanga, and Matt Stichter.

Third, different groups provided me with financial support during the writing of the dissertation, and then the book. For the dissertation, financial support came from The Bowling Green State University Philosophy Department, the Earhart Foundation, and the Charlotte and Walter Kohler Charitable Trust, who support the Kohler Fellows program at the Howard Center for Family, Religion, and Society. For the book, financial support came from California State University-Sacramento.

Fourth, a number of colleagues here at CSUS have given various types of support in writing and bringing the book to print. For example, Tom Pyne, Chris Bellon, Randy Mayes, and Scott Merlino all gave constructive feedback when I read a paper which became the basis for Chapter 2 and 3 in the spring of 2006. Jeremy Garrett was instrumental in encouraging me and helping me to contact Springer with the manuscript.

Fifth, the kind and professional people working in or with Springer's Philosophy and Medicine Series have been a delight to work with: Aaron Hinkley, Lisa Rasmussen, Marion Wagenaar, and Indumadhi Srinivasan. Also, Chris Tollefsen and another anonymous reviewer at Springer made many (!) excellent (!!) comments challenging me to rethink and expand on various parts of the argument.

Sixth, a number of journals kindly granted permission to reprint material that originally appeared in their pages. "Capacities, Hierarchies, and the Moral Status of Normal Human Infants and Fetuses." *Journal of Value Inquiry* 43: 479–492 (December 2009). "A Qualified Endorsement of Embryonic Stem-Cell Research, Based on Two Widely Shared Beliefs about the Brain-Diseased Patients Such Research Might Benefit." *Journal of Medical Ethics* 34: 563–567 (July 2008). "Not Every Cell is Sacred: A Reply to Charo." *Bioethics* 20(3): 146–157 (June 2006). "Human Embryos in the Original Position?" *Journal of Medicine and Philosophy* 30(3): 285–304 (June 2005). Of course, when thanking journals, let me also thank all of the anonymous reviewers out there who read, commented on, rejected, and sometimes accepted, bits and pieces of this book when it was under review as journal submissions: you know who you are, even though I do not!

Seventh, a number of publishers agreed to let me reprint longer portions of the following pieces:

Jeff McMahan, *The Ethics of Killing: Problems at the Margins of Life*; copyright © 2002: Oxford University Press; reproduced here by permission of Oxford University Press, Inc.

Michael Tooley, *Abortion and Infanticide*; copyright © 1983: Oxford University Press; reproduced here by permission of Oxford University Press.

The President's Council on Bioethics, *Controversies in the Determination of Death: A White Paper by the President's Council on Bioethics*; copyright © 2009: The President's Council on Bioethics; reproduced here by kind permission of The President's Council on Bioethics.

Reprinted from The Lancet, volume 358 (December 15, 2001), Pim van Lommel, Ruud van Wees, Vincent Meyers, and Ingrid Elfferich, "Near-death experience in survivors of cardiac arrest: a prospective study in the Netherlands," pages 2039–45; copyright © 2001, with permission from Elsevier.

Alfonso Gomez-Lobo, "Sortals and Human Beginnings"; copyright © 2004: Alfonso Gomez-Lobo; reproduced here by kind permission of the author. http://ontology.buffalo.edu/medicine_and_metaphysics/Gomez-Lobo.doc

Eighth, as this book was in its final stages of production, Martha Nussbaum graciously pointed out to me that her views (which I interact with in Chapter 3) had changed in ways that I had not been aware of, thereby saving me from a number of potentially embarrassing errors. While this acknowledgment should not be construed as her endorsement of my argument, I am grateful to her for giving me the opportunity to significantly improve that part of my argument prior to publication.

Finally, my immediate and extended family gave me various kinds of valuable support throughout the process of writing the material that has now become this book: my in-laws Phillip, JoAnne, Ian, Sean, and Nick McDaniel; my siblings Frank and Elizabeth; my parents Ruth and Frank; my children Hannah, Grace, Bethany, Joshua, and Mary; and last of all, most of all, and with my all, I thank my wife, Heather. There are things that philosophers simply cannot do with words; expressing adequately my thanks to you, Heather, is one of them.

Contents

1 You Are Not What You Think: Capacities, Human Organisms, and Persons 1
1 The Adventure of Ronald Reagan's Brain 1
2 What Are Humans? 8
3 What Is Serious Moral Status? 10
4 What Are Typical Human Capacities? 17
5 What Are Persons? 26
6 What Are We? 29

2 Anything You Can Do, I Can Do Also: Humans, Our Capacities, and the Powers We Share 35
1 How to Compare Capacities Between Individuals 35
2 A Temporary Change Argument About What We Are 37
3 The Capacities of Undeveloped Human Organisms 42
4 The Capacities of Damaged and Disabled Human Organisms 55

3 The Only Game in Town: Why Capacities Must Matter Morally 65
1 A Temporary Change Argument About What Matters Morally 65
2 Moral Status and the Past 68
3 Moral Status and the Future 70
4 Why Not Stop at the First-Order Capacity? 73
5 Actual, Continuing Subjects of Experience 79
6 Capacities and the Original Position 85
7 Capacities and the Capabilities Approach 94

4 Little People: Higher-Order Capacities and the Argument from Potential 107
1 The Dreaded Argument from Potential 107
2 Not Every Cell Is Sacred 110
3 Potential Presidents and Potential Persons 131

5 Not Just Damaged Goods: Higher-Order Capacities and the Argument from Marginal Cases 143
1 The Dreaded Argument from Marginal Cases 143

2 Tooley's Cat, Boonin's Spider, McMahan's Dog, and Balaam's Ass 146
3 How Not to Be a Speciesist 163

6 Old Objections and New Directions: Capacities and Moral Status at the Very Borders of Human Life 165
1 Does the Temporary Change Argument Prove Too Much? 165
2 The Corpse Problem 169
3 Solving the Corpse Problem 174
4 Are Active Capacities Preferable to Passive? 184
5 Drawing Lines Near Altered Nuclear Transfer and Anencephaly 194

Bibliography 201

Index 205

Introduction

The thesis of this book is that if something is human, it has the sort of moral status that you and I each have—what I shall henceforth call "serious" moral status. The reason for labeling it "serious" is to explicitly distinguish it from the many real yet lower-grade sorts of moral status that other things in the universe have, such as works of art, natural landscapes, plants, trees, and at least some non-human animals. The reason for labeling it "serious" instead of "human" is that there could be other things besides humans that have the sort of moral status that you and I each have. Or so I argue.

I take it for granted that you and I each have serious moral status. This assumption is not something I argue for in this book. But this assumption is not controversial. Each of us firmly believes that we have serious moral status. You believe, for example, that other people owe you a certain amount of respect, and that they should not try to harm you unless they have very good reasons to do so.

Serious moral status has many dimensions or aspects besides the two just mentioned, which focus on respect and harm. Perhaps certain natural rights are a part of serious moral status, such as the rights to life, liberty, and the pursuit of happiness. Perhaps serious moral status prescribes a certain sort of aesthetic response on the part of those who encounter it in others. Like love, serious moral status is a many-splendored thing. The nature of serious moral status—its precise shape, structure, and content—is something open to debate. But the nature of serious moral status, like the reality of your moral status and mine, is not something I argue for in this book.

What I argue for in this book is a pair of claims, which focus, not on the *reality* or *nature* of serious moral status, but on the *basis* of serious moral status—that is, on the features about us in virtue of which we possess serious moral status. The thesis of the book, recall, is that if something is human, it has serious moral status. Is being human, then, the basis of serious moral status? Well, yes and no. I argue that being human involves possessing a feature that is itself a basis of serious moral status, in anything that possesses it, whether its possessor is human or not. I argue that the feature of us humans, in virtue of which we have serious moral status, is the possession of certain *capacities*.

The move towards capacities, in the context of discussions about moral status, is both common and controversial. It is common because many current debates about the moral status of things—whether the things in question are nonhuman animals, human fetuses, or what have you—eventually migrate towards a discussion of the capacities of the things in question—for example, the capacity to feel pain, the capacity for self-consciousness, and the capacity to think rationally. However, the move towards capacities is controversial because, more often than not, this move is an indication—indeed, a bright red flag—that certain humans are about to be assigned a different moral status than the sort of moral status you and I each have. After all, if a human's capacities are the basis of its moral status, how could a human having lesser capacities than you and I have the same "serious" moral status as you and I? For this reason, the move towards capacities is strongly resisted in some quarters—and eagerly welcomed in others.

If the main argument of this book is correct, then resisting (or welcoming) capacities for this reason is premature. For this main argument consists in a pair of claims, which, taken together, entail that all human beings have the same sort of moral status that you and I each have:

1. If something is human, it has a set of typical human capacities.
2. If something has a set of typical human capacities, it has serious moral status.
   Therefore,
3. If something is human, it has serious moral status.

This argument is sound, I argue, because typical human capacities include both what might be called "active" capacities and "passive" capacities, and also include both what might be called "lower-order" capacities and "higher-order" capacities. Although both distinctions are somewhat rough, I have an active capacity to raise my arm on purpose, a passive capacity to feel pain when pricked, lower-order capacities to do each of these things right now, and higher-order capacities to do each of these things even when I am relatively "incapacitated": for example, when I am temporarily comatose. Most writers who focus on the moral relevance of capacities tend to ignore a thing's higher-order capacities, or its passive capacities, or both. But I argue that all of a thing's capacities—whether active or passive, whether higher-order or lower-order—are relevant to its moral status. Allowing passive higher-order capacities to be relevant to an entity's moral status solves a number of problems that are otherwise very difficult to solve. But, as we shall see, allowing passive higher-order capacities to be relevant in this way also generates many new problems that other accounts do not face.

So, then, this book brings together a discussion in ethics, over the existence, nature, and types of moral status, with a discussion in metaphysics, over the existence, nature, and types of capacities. I argue that certain metaphysical distinctions among capacities are helpful for answering questions in the moral arena, and that the position I advance is better at solving certain problems than other positions.

This book employs a certain sort of controversial methodology, and reaches certain sorts of controversial conclusions. The methodology I employ relies upon our intuitions: in particular, it relies not only upon our intuitions about familiar cases, but also upon our intuitions about unusual—some would say metaphysically bizarre—thought experiments from time to time. Some of the hypothetical cases in my argument involve machines that can instantly duplicate a human body, futuristic brain surgery that cannot currently be practiced in modern hospitals (at least not legally), non-human aliens, and causal overdetermination. In other words, I am working within a stream of contemporary philosophy that imitates the work of authors like Jeff McMahan, Derek Parfit, and Michael Tooley.[1] The methodology I employ is also controversial because it relies upon the work of contemporary moral and political philosophers like John Rawls and Martha Nussbaum. This, by itself, is not especially controversial, since these philosophers are widely discussed already. What makes reliance upon Rawls and Nussbaum controversial in my case is the fact that I use their work in tandem with the unusual thought experiments, and the fact that I use their work to reach the controversial conclusions.

The conclusions I reach commit me to the idea that human fetuses and embryos have serious moral status. They also commit me to the idea that serious moral status is possessed by humans in a permanent vegetative state, humans suffering from terrible brain diseases, and humans born with terrible genetic disorders. In other words, I am reaching conclusions that are just as "conservative" and "pro-life" as those reached by many Roman Catholic philosophers such as Pope John Paul II. Indeed, in certain places, and for reasons that will become clear as the argument progresses, I reach conclusions that seem to be *more* conservative than some leading Roman Catholic writers in the natural law tradition. However, no theological premises are involved in any part of my argument. This is not only a book that a pro-life Catholic can give to an atheist in order to explain why the Catholic position is correct as far as it goes; it is also a book that a pro-life atheist can give to a Catholic in order to explain why the position of some Catholic authors does not go quite far enough.

This, then, is a book that begins with certain intuitions we all have about ourselves, works through a few metaphysical distinctions, a few thought experiments, and a few contemporary moral philosophers, and ends up with substantive, and controversial, conclusions about the moral status of humans at the beginning and ending stages of life. If the argument of this book is sound, this will have important implications for the proper way of framing a number of debates in the context of biomedical ethics. For if all human organisms really do have serious moral status, whether or not they are very undeveloped, diseased, or damaged, then this must have some impact on our ongoing moral and political debates about the proper treatment of such organisms at various stages of their biological lives.

The chapters of the book are structured as follows. Chapter 1 explains the concepts of the main argument in more detail, and explains why personal pronouns such as "you" and personal names such as "Ronald Reagan" are applied to human

[1] See McMahan, J. *The Ethics of Killing: Problems at the Margins of Life*. Oxford: Oxford University Press (2002); Parfit, D. *Reasons and Persons*. Oxford: Clarendon Press (1984); Tooley, M. *Abortion and Infanticide*. Oxford: Clarendon Press (1983).

organisms throughout the book. Chapter 2 defends the first step of the main argument by focusing on human organisms that undergo temporary changes involving "incapacitation" of one sort or another. Chapter 3 defends the second step of the main argument, partly by focusing on the same sorts of "incapacitation" cases that appeared in Chapter 2, and partly by focusing on moral arguments that emerge from Rawls and Nussbaum. Chapter 4 and 5 relate the main argument to two controversial arguments in contemporary applied ethics: the Argument From Potential, which focuses on normal human organisms at the beginning stages of life (such as human infants, fetuses, and embryos), and the Argument From Marginal Cases, which focuses on abnormal human organisms (such as human organisms that are disabled, diseased, or genetically deficient). Finally, Chapter 6 considers a number of lingering objections, focusing especially on those objections related to my methodology, my treatment of the line between life and death, and my treatment of the line between defective humans and nonhuman entities.

Chapter 1
You Are Not What You Think: Capacities, Human Organisms, and Persons

1 The Adventure of Ronald Reagan's Brain

To grasp the basic argument of this book, it helps to start with a vivid example.

When former president Ronald Reagan died from Alzheimer's disease, it provided an unusual opportunity for continuing the debate about embryonic stem-cell research (ESCR). Proponents of such research argued that it should be aggressively funded by the federal government, since it might lead to the discovery of cures for diseases like Alzheimer's. For example, Ron Reagan Jr., the former president's son, gave a speech at the 2004 Democratic National Convention in which he argued that even those whose opposition to federally funding ESCR is "well-meaning and sincere" only have "an article of faith" to base their opposition upon. Although "they are entitled to" this article of faith, Reagan Jr. continued, "it does not follow that the theology of a few should be allowed to forestall the health and well-being of the many".[1] Opponents of ESCR, on the other hand, argued that former president Reagan would have been opposed to it, since this research typically involves the destruction of living human organisms, and he believed that every human being—here understood as a human organism—possessed a sacred and inviolable dignity from its conception until its natural death. For example, Michael Reagan, another son of the former president, wrote an article titled "I'm With My Dad On Stem Cell Research," in which he wrote:

> I'm getting a little tired of the media's insistence on reporting that the Reagan "family" is in favor of stem cell research, when the truth is that two members of the family have been long time foes of this process of manufacturing human beings—my dad, Ronald Reagan during his lifetime, and me.[2]

Whether ESCR could lead to cures for degenerative brain diseases is disputed. When Ron Reagan Jr. gave his speech at the DNC, he invited his listeners to imagine a day when ESCR had provided a cure for Parkinson's disease:

---

[1] Reagan (2004).

[2] Reagan (2007).

R. DiSilvestro, *Human Capacities and Moral Status*, Philosophy and Medicine 108,
DOI 10.1007/978-90-481-8537-5_1,

> Let's say that ten or so years from now you are diagnosed with Parkinson's disease. . .Now, imagine going to a doctor who, instead of prescribing drugs, takes a few skin cells from your arm. The nucleus of one of your cells is placed into a donor egg whose own nucleus has been removed. A bit of chemical or electrical stimulation will encourage your cell's nucleus to begin dividing, creating new cells which will then be placed into a tissue culture. Those cells will generate embryonic stem cells containing only your DNA, thereby eliminating the risk of tissue rejection. These stem cells are then driven to become the very neural cells that are defective in Parkinson's patients. And finally, those cells —with your DNA—are injected into your brain where they will replace the faulty cells whose failure to produce adequate dopamine led to the Parkinson's disease in the first place. In other words, you're cured.

However, when Michael Reagan considered the idea that ESCR can lead to a cure of Alzheimer's disease in his article, he called the idea a "widely promoted and thoroughly discredited junk science argument" that merely "helps generate public support for the biotech political agenda."

Assume for a moment, what may after all be true, that human ESCR will one day lead to a regenerative cure for brain diseases like Parkinson's and Alzheimer's. This sort of cure, let us assume, could work in the following restorative way: if the relevant stem-cell tissues were injected into a certain region of a patient's brain after a disease had done its characteristic damage to that region, then that region of the brain would become capable of being restored to the state it was in before the disease had done its characteristic damage.

Now consider the following pair of hypothetical cases:

> Case A: On the day Reagan receives the diagnosis, his doctors give him the regenerative cure. Reagan's mental status never deteriorates further. Some damage done by the disease is repaired: for example, damaged brain tissues involved in certain functions are restored, and the relevant functions improve. But some damage remains: for example, damaged brain tissues which supported the memory of his famous speech at the Berlin Wall are not restored, and that memory is never recovered.

> Case B: Same as Case A, only this time, all the damage done by the disease gets repaired: even the tissues which supported the memory of his famous speech at the Berlin Wall are restored.

To explain the difference between these cases, we must appeal to something beyond mere stem-cell treatment. Here is one such appeal: imagine that the doctors have at their disposal a "brain scanner" that can do two things. First, using electromagnetic imaging, the scanner can record a three-dimensional "snapshot" of the exact structure of a given brain region at a given time—call this "scanning out". Second, using nanotechnology, the scanner can reproduce that exact structure in the same brain region, at a different time, as long as a certain amount of basic brain structure is already present—call this "scanning in". If Reagan's doctors had done a "scan-out" of each of Reagan's brain regions before the disease had even begun to do its damage, they could do a "scan-in" after his stem-cell injection to reproduce the exact structure of the tissues supporting his memory of the Berlin Wall speech.

Next, consider another pair of cases:

> Case C: On the day Reagan receives the diagnosis, his doctors do not have stem cell therapy or the ability to "scan-in", but they can "scan-out". So they take a snapshot of Reagan's

brain, but must watch helplessly as Reagan's mental status deteriorates significantly over the following months and years. The disease causes him to lose, among other things, all his memories. However, when the stem cell therapy becomes available, Reagan's doctors use it, and all the significant structural damage done by the disease is repaired. And when the "scan-in" technology becomes available, the doctors use it to regain all the lost memories.

Case D: Same as Case C, only this time, assume that Reagan had a brain disease even more destructive than Alzheimer's. Assume that the regenerative cure was put in the hands of Reagan's doctors before Reagan's biological death, but after his disease had destroyed those brain structures that supported his distinctive memories, beliefs, and desires, and after it had also destroyed those brain structures of Reagan's that supported perception, conscious self-awareness, and all other mental states. Assume that, although a living human organism remained, there were no experiences at all being had by that organism.

Consider what the recovery would look like in Case D. Reagan's doctors inject the relevant stem-cell tissues into the relevant region of Reagan's brain. That region of the brain becomes capable of being restored to the state it was in before the disease had done its characteristic damage. In a way similar to the way a starfish starts growing back a limb that has been removed, Reagan's brain starts growing back the structures that had been destroyed by his brain disease. However, even though the injection makes a region of Reagan's brain capable of being restored to its pre-disease state, there is no guarantee, without the brain scanner, that this region will be restored to exactly the same structure it had before the disease took its toll. For even if we assume that Reagan's brain develops the characteristic structures necessary for supporting perception, conscious self-awareness, and other mental states, the exact details of those characteristic structures might be very different than the structures that supported the distinctive memories, beliefs, and desires that most people would associate with Reagan.

The brain scanner solves this problem. Since they had already "scanned-out" Reagan's brain structure on a certain agreed-upon day before the brain disease had much of a chance to run its destructive course, they are able to reproduce that exact structure into Reagan's regenerated brain, after the brain has been given the stem-cell injection and has been able to develop the minimum structure needed for the "scan-in". The result is that the brain structures supporting beliefs, desires, and so on, on the day of the "scan-in", would be exactly the same as the brain structures that supported beliefs, desires, and so on, on the day of the original "scan-out".

Although each of these cases is dramatic its own way, let us focus on Case D, calling it "the adventure of Reagan's brain" to emphasize the truly stunning brain recovery it represents. Some opponents and some proponents of ESCR would agree with the following two claims about this adventure:

(1) Reagan himself survives each stage of this adventure.
(2) Reagan retains his moral status during each stage of this adventure.

Since I use these claims to argue for some controversial claims below, I need to say a bit more about what they mean and why they would be attractive to both some opponents and some proponents of ESCR.

The meaning of (1) is straightforward. Reagan was there at the beginning, before the disease began its work. He was still there in the middle, after the disease had taken its terrible toll. And he was still there at the end, after the injection and scan-in. Reagan persisted throughout the adventure. It was, in fact, *his* adventure, and the different parts of the adventure, such as the psychological deterioration and recovery, were *his* psychological deterioration and *his* recovery.

Some people, regardless of their views about ESCR, would agree with (1). Such people believe that a debilitating brain disease makes a diseased patient worse off, and that, if a regenerative cure became available for such a disease, this cure, if used, would make the diseased patient better off. Such people believe that it is the patient himself, and not some other thing, made worse off by a brain disease, and that it is the patient himself, and not some other thing, made better off by a regenerative cure. Such people believe that a patient persists through the ups and downs of a brain disease and cure, even when the downs go as low as in Cases C and D.

Belief in (1) can be substantiated by at least three arguments. First, it is at least tempting to accept the idea that Reagan himself exists *after* the adventure, since the individual after the adventure has the same living body, and the same memories, beliefs, and desires, as the individual *before* the adventure (namely, Reagan). But this idea, when combined with the attractive assumption that persons cannot have temporal gaps, leads directly to the conclusion that Reagan himself exists in the *middle* of the adventure: in other words, Reagan persists through each stage of the adventure.

Second, if one asks why we should accept the idea that Reagan himself exists after the adventure, the following answer is available. If you discovered that the sort of psychological deterioration and recovery described in Cases A–D were about to happen to you (or, to put the matter as neutrally as possible, to the biological organism with which you are currently associated), you would be concerned about the pleasures and pains of the individual that emerges from the recovery in all of them, not just with the sort of "sympathetic" concern that you have for the pleasures and pains of another, but rather with the sort of "egoistic" concern that you have about your own future pleasures and pains. This egoistic concern is both psychologically spontaneous and difficult to shake. If a given view coheres well with your egoistic concern, that is a reason for accepting the view. But the only view that coheres well with your egoistic concern is the view that the individual that emerges from the recovery is, in fact, you. And what's true for you in your hypothetical adventure is also true for Reagan in his.

Third, consider a number of hypothetical mini-adventures which we are confident of Reagan surviving. If a mini-adventure only involves Reagan losing his memory of his speech at the Berlin Wall, but nothing more, before the injection and scan-in—as in Case B—we should think Reagan survives it. The same could be said if the mini-adventure involved losing and regaining the memory of his wedding. Reagan would survive all such "memorial" mini-adventures, even if they took place one right after the other. But then it seems Reagan would survive them, even if they took place, so to speak, "all at once": even if a mini-adventure involved Reagan losing all his memories, before the injection and scan-in—as in Case C—we should still think Reagan survives it. Next, what is true for memories is true for other mental powers: even if a mini-adventure involved Reagan losing all his perceptual abilities, before the injection and scan-in, we should still think Reagan survives it. Reagan would survive all such "mental" mini-adventures, even if they took place one right after the other. But then it seems Reagan would survive them, even if they took place, so to speak, "all at once": even if an adventure involved Reagan losing all his mental powers, before the injection and scan-in—as in Case D—we should still think Reagan survives it.

So, then, some opponents and some proponents of ESCR agree with (1): that Reagan himself survives each stage of the adventure in Case D. Not everyone agrees with (1), as we shall see. But (1) is a widely shared belief that can be backed up with arguments.

The meaning of claim (2)—"Reagan retains his moral status during each stage of this adventure"—can be clarified by a definition and an example. According to Mary Anne Warren's definition,

> To have moral status is to be morally considerable, or to have moral standing. It is to be an entity towards which moral agents have, or can have, moral obligations. If an entity has moral status, then we may not treat it in just any way we please; we are morally obliged to give weight in our deliberations to its needs, interests, or well-being. Furthermore, we are morally obliged to do this not merely because protecting it may benefit ourselves or other persons, but because its needs have moral importance in their own right.[3]

Although one position says that there is only one kind of moral status, and that only humans can have it, a better position says that many kinds of individuals can have moral status, and that different kinds of individuals—Reagan and his dog, let's say—can have different kinds of moral status (Reagan has his moral status, his dog has another).

Claim (2) is a claim about Reagan retaining his characteristic moral status. To clarify (2), consider the following amendment of Case D. If one of Reagan's enemies sneaks into his hospital room, the night before the stem-cell injection, and ends his life by giving him a painless injection of some drug that stops his heart from beating, then this enemy has violated Reagan's moral status. This is a wrong-making feature of her action. Her action would have possessed additional wrong-making features if she had killed Reagan painfully, before the brain disease had begun, in the presence of Reagan's family and friends. But this will not distract us from the fact that, by ending Reagan's life the night before his stem-cell injection, Reagan's enemy violated Reagan's moral status. But in order for her to violate Reagan's moral status by her action, it seems that Reagan must possess his moral status at the time of her action.

Some people, regardless of their views about ESCR, would agree with (2). Such people believe that brain-diseased patients who stand to benefit from regenerative cures should be treated as objects of moral consideration and concern. Such people believe that a patient does not somehow lose his moral status just because of

[3] Warren (1997, p. 3).

a temporary damage to a part of his brain. Put roughly and in terms of "rights," the idea is that even if a patient temporarily loses some of his rights (e.g., the right to drive a car) because of temporary brain deterioration, he still retains a number of rights (e.g. the right to life) even during the time of temporary brain deterioration.

Belief in (2) can be substantiated by at least two arguments. First, if you discovered that the sorts of psychological deterioration and recovery described in Cases A–D were about to happen to you, you would anticipate retaining your characteristic moral status throughout all of them. This anticipation is both psychologically spontaneous and difficult to shake. If a given view coheres well with this anticipation, that is a reason for accepting the view. But the only view that coheres well with this anticipation is the view that you really do retain your characteristic moral status throughout your adventure. And what's true for you in your hypothetical adventure is also true for Reagan in his.

Second, the idea that Reagan retains his moral status through each stage of this adventure can be supported by considering the same hypothetical mini-adventures considered earlier. We are confident that Reagan would retain his moral status through the mini-adventure described in Case B. But since Reagan would retain his moral status through all such "memorial" mini-adventures if they took place sequentially, so too Reagan would retain his moral status through a mini-adventure involving the loss of all his memories, as in Case C. Once again, what is true for memories is true for other mental powers: since Reagan would retain his moral status throughout all such "mental" mini-adventures if they took place sequentially, so too Reagan would retain his moral status through an adventure involving the loss of all his mental powers, as in Case D.

So, then, some opponents and some proponents of ESCR agree with (2): that Reagan retains his moral status during each stage of the adventure in Case D. Not everyone agrees with (2), as we shall see. But (2) is a widely shared belief that can be backed up with arguments.

I would now like to explain why anyone who accepts (1) and (2) must accept the following claim:

> (3) Any theory of personal identity or moral status that cannot accommodate (1) and (2) should be rejected.

Some theories of personal identity cannot accommodate (1). They entail that Reagan literally ceased to exist at some point in the adventure.[4] All that really survives each stage of the adventure, according to these theories of personal identity, is a living human organism.[5] A living human organism persists through the adventure, but this does not guarantee that Reagan persists through the adventure. For Reagan is not a living human organism. According to many theories of personal

---

[4] An excellent introduction to different approaches to personal identity, with special attention to their relevance to bioethics, is DeGrazia (2005).

[5] A recent philosophical defense of the view that persons are human organisms is Olson (1997). A recent philosophical critique of this view is found in McMahan (2002, pp. 3–94).

identity, Reagan is something else. On some of these theories, the living human organism in the middle of the adventure is neither Reagan nor anyone else: it is just an anonymous, impersonal nobody.

Now, some theories of personal identity can simultaneously deny that Reagan is a living human organism, and yet affirm that Reagan persists through the adventures. However, according to many theories of personal identity that deny that Reagan is a living human organism, Reagan literally ceases to exist by some point in the progression of the brain disease, and thus (1) is false.

According to (3), theories of personal identity that cannot accommodate (1) should be rejected. And those who accept (1) must accept (3). Just as we all must say "if a given theory of personal identity cannot accommodate the claim that I survived the last 24 hours, 8 of which were spent in a deep, dreamless sleep, then this is a reason for rejecting the theory," someone who accepts (1) must say "if a given theory of personal identity cannot accommodate the claim that it was Reagan himself that survived this adventure, then this is a reason for rejecting the theory."

Some theories of moral status cannot accommodate (2). Many such theories deny (2) because they deny (1). "After all," such theories say, "although Reagan's enemy killed a living human organism, she did not kill Reagan. Reagan was not there to be killed." But it is possible to believe (1) and still deny (2). For example, some theories say that Reagan's moral status is the sort of thing that fluctuates over the course of his biological life depending on what else is true of him: when his upper brain is intact, before the disease runs its course, Reagan has moral status, but when the disease (temporarily) obliterates the structure of his upper brain, Reagan's moral status is (temporarily) obliterated as well.

According to (3), such a theory of moral status should be rejected. And those who accept (2) must accept (3). Just as we all must say "if a given theory of moral status cannot accommodate the claim that I myself retained my moral status during last night's deep, dreamless sleep, then this is a reason for rejecting the theory," someone who accepts (2) must say "if a given theory of moral status cannot accommodate the claim that Reagan himself retains his moral status throughout this adventure, then this is a reason for rejecting the theory."

If you disagree with (1) or (2), I invite you to consider them as elements in what I take to be a better picture of both personal identity and moral status than the pictures often presented by contemporary philosophers who write on these topics. (1) and (2) are my starting points, and the rest of this book attempts to both defend these starting points from rival accounts of personal identity and moral status, and to develop from these starting points an alternative approach to personal identity and moral status. When this is done, I believe that we will be led to a view that, on the one hand, identifies persons like Ronald Reagan with human organisms, yet, on the other hand, also allows for the possibility that persons like Ronald Reagan could exist in a disembodied state. We will also be led to a view that, on the one hand, allows for all human organisms to have serious moral status, yet, on the other hand, also allows for the possibility of non-human entities to have the same serious moral status as human organisms.

The main argument of this book consists in a pair of claims, which, taken together, entail that all human beings have the same sort of moral status that you and I each have:

1. If something is human, it has a set of typical human capacities.
2. If something has a set of typical human capacities, it has serious moral status.
Therefore,
3. If something is human, it has serious moral status.

To defend this argument, it is important to clarify what is meant by "human," "serious moral status," and "a set of typical human capacities." The next three sections do just this.

2 What Are Humans?

The term "human" is sometimes used ostensively. When asked what beings in the world are human, most of us would begin our answer simply by pointing and by saying, "why, those things, over there, are human. . .this thing, right here, is human. . . I myself, for example, am human. . .I'm a human being." But ostension only takes us so far, because eventually different people with different understandings about what property "human" refers to will point to the same thing and disagree about whether it is human.

When we attempt to go beyond mere ostension, we quickly encounter a number of different ways, or families of ways, for using the term "human". The term "human" sometimes refers to being an individual member of the species Homo sapiens, or being an individual member of the species Homo sapiens with a certain sort of mental life, or being a rational animal, or being made in God's image. This term sometimes gets used in discussions of the property of possessing basic "human" rights, such as the right to life, liberty, and the pursuit of happiness. But it also sometimes gets used in discussions of a property focused on by Immanuel Kant, of being an "end in itself" that deserves the respect of all rational agents. Furthermore, the term "human" sometimes refers to the properties we use as *evidence* for classifying an individual as a member of the species Homo sapiens. Thus "human" sometimes refers to a broadly phenotypic property, such as the property of being a featherless biped, or the property of having phenotypes "like those things, over there" (using ostension to designate the comparison class). It sometimes refers to a broadly genotypic property, such as the property of possessing a certain number of chromosomes, or the property of having a certain sort of DNA, or the property of having genotypes "like those things over there" (again, using ostension to designate the comparison class). It sometimes refers to a broadly genealogical property (provided that one of the other senses of "human" is already in place), such as the property of being an ancestor of human beings, or the property of being a descendant of human beings, or the property of being a relative of human beings, or the property of being able to produce viable offspring with human beings.

In the main argument of this book, an entity is "human" if it is an organism that has possessed, at some time in its past history, the same basic genotype as you and I possess right now. More precisely: an entity is human if it is an organism that has possessed, at some time in its past history, a structure made up out of one or more cells that have the same basic genotype as your cells and my cells have right now. Cells have the same basic genotype as yours and mine just in case they have DNA of the sort that you and I have even though chimpanzees do not. It is difficult to give a more precise characterization than this. But a creature with ZNA instead of DNA in its cells—where ZNA is a structure radically different than DNA, not having a double helix structure, not even having base pairs, and so on—would not have the same basic genotypes as us. Consider an organism whose cells have ZNA instead of DNA in them, yet whose phenotypes are indistinguishable from the phenotypes of Hillary Clinton: the organism looks and behaves just like her, and even has experiences just like hers (this organism knows the answer to the question "What is it like to be Hillary Clinton?"). Hillary Clinton and this organism have different basic genotypes; she is human, this organism is not. But Hillary Clinton and Barack Obama have the same basic genotype; they are both human.

So, then, an entity is "human" if it is an organism that has possessed, at some time in its past history, a structure made up out of one or more cells that have the same basic genotype as your cells and my cells have right now. The phrase "is an organism" is designed to avoid at least two sorts of obvious problems. First, assume Obama dies, and there is now a corpse to dispose of. This corpse possesses a structure made up out of one or more cells that have the same basic genotype as your cells and my cells have right now. My definition of "human," however, would entail that this corpse is not human. For it is not true that this corpse is an organism. This is, I believe, a desirable result, since ordinary language often suggests that corpses are only "human" in something like a derivative sense. Just as an archaeologist might call a piece of wood a "human" artifact because of its previous association with human organisms, so too she might call a set of bones "human" remains because they previously composed, or partly composed, a human organism. Since the word "human," as I am using it, should not be understood to refer to corpses, I hope that no one will construe my main argument as a defense of the view that corpses have serious moral status.

The example of the archaeologist who discovers a set of bones alerts us to the second obvious problem that the phrase "is an organism" is designed to avoid. Even while he is alive, Obama has a skeleton. This skeleton possesses a structure made up out of one or more cells that have the same basic genotype as your cells and my cells have right now. My definition of "human," however, would entail that Obama's skeleton is not human. For it is not true that this skeleton is an organism. This, too, I believe, is a desirable result, since ordinary language strongly suggests that mere parts of human organisms are only "human" in something like a derivative sense. Since the word "human," as I am using it, should not be understood to refer to mere parts of human organisms, I hope that no one will construe my main argument as a defense of the view that such mere parts have serious moral status.

Once again, then, an entity is "human" if it is an organism that has possessed, at some time in its past history, a structure made up out of one or more cells that have the same basic genotype as your cells and my cells have right now. The phrase "at some time in its past history" is designed to handle at least two sorts of thought-experiments. First, assume Hillary Clinton happens to be an immaterial soul that can exist disembodied after her biological death as a human organism. My definition of "human" would entail that she is still human during this disembodied phase of her existence. For even when she is disembodied, it is still true that she had possessed, at some time in her past history, the relevant sort of basic genotype. This, I believe, is a desirable result, since it allows for Hillary Clinton to continue being human even after her biological death as a human organism. Second, assume Hillary Clinton happens to be an immaterial soul that can exist disembodied *before* her biological life as a human organism. My definition of "human" would entail that she is not human during this disembodied phase of her existence. For when she is disembodied, it is not true that she had possessed, at some time in her past history, the relevant sort of cell structure. This too, I believe, is a desirable result, since it does not make Hillary Clinton human before her biological life as a human organism.[6]

3 What Is Serious Moral Status?

The phrase "moral status" has at least five distinct, but closely related, meanings. First, talk about the moral status of something is often simply a shorthand way of referring to its morally salient features, or those of its features that are prominent from a moral point of view. For example, when we debate whether some action or policy is morally required, or morally permissible, or morally impermissible, we are

[6] I am assuming, of course, that Hillary Clinton would continue to be an organism even if she became disembodied. Some might reject this assumption on the grounds that an entity needs to have some physical *organs* to be an organism, and since a disembodied soul does not have physical organs, it follows that a disembodied soul cannot be an organism. I have two replies to this objection. First, it is not obvious to me that an entity needs to have some physical organs to be an organism. I am attracted to the view that what an entity needs, in order to be an organism, is a certain sort of *organization* of its parts (if it has any), properties, and processes. Thus, as long as the entity has the relevant sort of organization, it could be a physical object, a computer program, an angel, or an immaterial soul. Second, if one wants to insist that disembodied souls cannot be (or remain) organisms once they become disembodied, then the definition of "human" offered in the text could be adjusted in either of the following ways: (1) "An entity is "human" if it is, *or was*, an organism *possessing* a structure made up out of one or more cells that have the same basic genotype as your cells and my cells have right now." This revised definition would still avoid making corpses count as humans, as long as one is willing to deny that a corpse *was* an organism. (2) "An entity is "human" if it is, or was, an *individual* possessing a structure made up out of one or more cells that have the same basic genotype as your cells and my cells have right now." This revised definition would still avoid making corpses count as humans, as long as one is willing to deny that a corpse *was* an individual.

debating the moral status (in this first sense) of that action or policy. Examples of this first sense of moral status are found in article titles like "The Moral Status of Abortion."[7]

Second, the phrase "moral status" is sometimes used to focus on the morally salient features of particular individuals in the world. For example, when we debate whether some entity has rights, or dignity, or intrinsic value, we are debating the moral status (in this second sense) of that entity. An example of this second sense of moral status is found in the following advertisement describing a recent philosophy conference:

> The notion of an entity's *moral status* is among the most fundamental and pervasive of our moral concepts. Current moral debates over stem cell research or therapeutic cloning, for example, revolve largely around questions concerning the moral status of embryos—the sort of intrinsic importance they may have (or lack), making them worthy (or unworthy) of certain kinds of moral consideration—just as debates over abortion or animal rights have long dealt with questions about the moral status of fetuses or of non-human animals. And the idea of the moral status of persons is central to many accounts of our own fundamental rights and obligations to one another, placing it at the very heart of ethical thought.[8]

This second concept of moral status is narrower than the first one, since it can only apply to concrete individuals in the world, and cannot apply to things like actions or policies. However, the claim that some item *has* moral status, in either the first sense or the second sense, is not very informative because it does not tell us what *sort* of moral status that item has. For example, the claim that some action has moral status does not tell us whether that action is morally required, or morally permissible, or morally impermissible, and the claim that human embryos have moral status does not tell us exactly what "the sort of intrinsic importance" is that they have "making them worthy. . .of certain kinds of moral consideration." The claim that something *has* moral status, in either this first or second sense, is rather like the claim that something *has* height: even if we know that something has height, we still do not know just how tall it is.

Third, there is Mary Anne Warren's concept of moral status, quoted earlier:

> To have moral status is to be morally considerable, or to have moral standing. It is to be an entity towards which moral agents have, or can have, moral obligations. If an entity has moral status, then we may not treat it in just any way we please; we are morally obliged to give weight in our deliberations to its needs, interests, or well-being. Furthermore, we are morally obliged to do this not merely because protecting it may benefit ourselves or other persons, but because its needs have moral importance in their own right.[9]

On Warren's gloss of the concept of moral status, the claim that some entity has moral status is very informative. For her, a thing has moral status just in case it has a cluster of other properties, such as being an entity towards which moral agents have

---

[7] Examples of this first use of moral status are found in Conn (2001), Kershnar (2001), Butler (1993) and Pojman (1992).

[8] Virginia Tech Department of Philosophy (2003).

[9] Warren (1997, p. 3).

moral obligations and being an entity whose needs have moral importance in their own right.

Fourth, Elizabeth Harman's concept of moral status builds upon the idea of a harm "mattering morally" as follows:

> A thing has moral status just in case harms to it matter morally...A harm to a being "matters morally" just in case there is a reason not to perform any action that would cause the harm and the reason exists simply in virtue of its being a harm to that thing, and simply in virtue of the badness of the harm for that thing... Some examples will help to illustrate this terminology. There are reasons not to harm both Alice and her car; but only Alice has moral status. Harms to Alice provide reasons against action simply in virtue of being harms to her. But harms to Alice's car provide reasons against action only in virtue of being harms to Alice; so these harms do not *matter morally* because the reason against action does not exist simply in virtue of the harm's being to that thing.[10]

Although more could be said about other senses of the phrase "moral status," enough has been said to introduce the concept of moral status that appears in the thesis of this book—what I call "serious moral status." Serious moral status is a species of both the first and second sorts of moral status considered above. It includes everything that Warren and Harman's concepts of moral status include. But it also includes much more as well. If something has serious moral status, then there is a strong moral presumption against harming it, a strong moral presumption against wronging it, and a strong moral presumption against even speaking ill of it, or "cursing" it in any way. If something has serious moral status, then it is owed respect, indeed owed justice, and there is a standing reason to benefit it whenever possible. Serious moral status is a place-holder for whatever morally salient features normal adult human persons, like you and I, possess.

One aspect of serious moral status sometimes goes by the name of "the right to life," but it can be characterized without the language of rights as follows: if an entity has this aspect of serious moral status, then there is a strong moral presumption against killing that entity. We all believe that other people are under a *presumption* not to kill us: the burden of proof is entirely on other people who wish to take our life to explain why they should be allowed to take it. We all believe that this presumption is *moral* (and not merely legal) in nature. And we all believe that this presumption is so *strong* that other people who wish to take our life must be able to give very convincing reasons for why they should be allowed to take it.

---

[10]Harman (2003, p. 174). See also Harman (2003), available online at http://hdl.handle.net/1721.1/17645. As she puts it in the dissertation, "If something is ever harmed, then it has moral status just in case we have reasons not to cause harms to it simply in virtue of the badness of the harms *for it*" (p. 15). Harman recognizes that her notion of moral status "is somewhat close to the intuitive meaning of "moral status"" even though it retains a somewhat stipulative character that reflects her own particular methodological purposes. For example, she explains her choice to include the phrase "if something is ever harmed" in her definition like this: "The antecedent "if something is ever harmed" in the explanation is necessary because if something is not ever harmed, then it vacuously satisfies the consequent: it is true that *all harms to it* have the right kind of reasons against them. For things that are never harmed, I advocate settling whether they have moral status by developing a substantive view of which things that are harmed have moral status" (p. 15).

The concept of a strong moral presumption against killing is an intuitive concept that can be used across various normative theories. This concept is intuitive because it is designed to capture two widely shared pre-theoretical intuitions, intuitions that guide the thinking of most people and that also guide the thinking of most philosophers who attempt to construct adequate theories about the morality of killing: first, that killing other normal adult human persons is prima facie morally wrong;[11] second, that this prima facie moral wrongness is strong enough to establish a presumption against killing other normal adult human persons. This concept is useful to various normative theories because admitting that there is a strong moral presumption against killing an individual still leaves three important questions open. First, what is the precise strength of this presumption? Second, what conditions might override this presumption? Third, what provides the basis of this presumption? The main argument of this essay does nothing to answer the first two of these important questions, but it does attempt to answer the third. For the second premise of the main argument is tantamount to the claim that one basis of serious moral status, and hence the strong moral presumption against being killed, is the possession of a set of typical human capacities.

Before explaining what sorts of capacities are the relevant ones, it is important to explain why I focus on an aspect of moral status that is both intuitive and theoretically neutral. I focus on an aspect of moral status that is intuitive because I want to bring on board as wide an audience of philosophers and non-philosophers as possible. The intuition that it is, on the face of it, morally wrong to kill normal adult human persons, is a stable and widely shared intuition. As Fred Feldman writes,

> One of the most widely accepted and intuitively plausible moral principles is "Thou shalt not kill." I take this to mean (or to imply) that it is morally wrong to kill people. It is hard to think of a moral principle with greater immediate credibility. Surely, if any moral principle is true, some version of this one is.[12]

The reason why I focus on an aspect of moral status that is theoretically neutral is that I do not want to prejudge certain important theoretical questions in moral philosophy. The concept of a strong moral presumption against killing, unlike the concepts of a right to life, or an absolute prohibition against killing, or the sanctity of life, does not raise questions about the existence of rights, or the absoluteness of prohibitions, or the meaning of "sanctity".

The connection between moral status and the morality of killing is widely recognized. For example, at the outset of his book *The Ethics of Killing: Problems at the Margins of Life,* Jeff McMahan claims that there are "four distinct categories into which we may sort most or all instances of killing for which there may be a reasonable justification," and the third category, which constitutes the primary focus of his book, consists of "cases in which the metaphysical or moral status of

---

[11] An action is prima facie morally wrong just in case the action, on the face of it, is morally wrong. Admitting that actions of a certain type are prima facie morally wrong does not commit one to any further thesis about what the wrong-making properties of that type of action are.

[12] Feldman (1992, p. 157).

the individual killed is uncertain or controversial."[13] To illustrate McMahan's point about controversy, there is controversy about the claim that there is a strong moral presumption against killing undeveloped human organisms. This claim is controversial partly because many arguments against it entail that there is not even a strong moral presumption against killing certain classes of developed human organisms. For example, consider the following argument:

(1) An entity must possess a mental life of a certain sort of richness in order for there to be a strong moral presumption against killing it.
(2) Undeveloped human organisms do not possess this sort of mental life.

Therefore,

(3) There is not a strong moral presumption against killing undeveloped human organisms.

The key premise in this argument is (1), but this premise, when combined with other highly plausible premises, such as

(4) Developed human organisms with certain forms of brain damage, dementia, or mental retardation do not possess this sort of mental life.

seems to entail that

(5) There is not a strong moral presumption against killing developed human organisms with certain forms of brain damage, dementia, or mental retardation.

McMahan and others are prepared to openly admit and discuss this sort of entailment. Consider the way he begins discussing his third category:

> Among those beings whose nature arguably entails a moral status inferior to our own are animals, human embryos and fetuses, newborn infants, anencephalic infants, congenitally severely retarded human beings, human beings who have suffered severe brain damage or dementia, and human beings who have become irreversibly comatose.[14]

The main argument of the present book, then, is set within the general context of contemporary disagreement and debate over the ethics of killing, and within the narrower context of a certain development within this debate, a development in which philosophers seriously entertain the idea that there is not a strong moral presumption against killing certain classes of adult human organisms because of their diminished moral status. I argue, against this idea, that all human organisms have serious moral status, which includes a strong moral presumption against being killed, and that all human organisms have this serious moral status because they have a set of typical human capacities.

---

[13] McMahan (2002, pp. vii–viii). The other three categories are (1) "cases in which killing would simply promote the greater good," (2) "cases in which an individual has *done* something that has lowered the moral barriers to harming him, or compromised his status as inviolable, or made him liable to action that might result in his death," and (3) "cases in which death would not be a harm to the individual but instead a benefit."

[14] McMahan (2002, p. vii).

Before moving on to explain what typical human capacities are, I need to clear up one possible confusion that might emerge from a comparison of this section (on the concept of serious moral status) and the previous section (on the concept of a human).

Since the concept of a human being I am working with is designed to make room for the possibility that a human being might survive its own biological death as a human organism, it is important to explain how the concept of serious moral status relates to this possibility. For it seems very plausible both that (a) an account of moral status should allow us to retain our moral status if we become disembodied, and (b) an account of the strong moral presumption against killing cannot allow there to be a strong moral presumption against killing us when we are disembodied. After all, if we become disembodied, we are already dead. And it is not possible to kill something that is already dead.

One way of handling the tension between (a) and (b) is to replace the concept of a strong moral presumption against killing with the following definition of the sort of moral status in question, which allows for human beings to retain the same more status after death even if they can become disembodied:

> For any X, X has moral status in the relevant sense if there is some type T, such that X is of type T, and there is a strong moral presumption against destroying anything of type T—where destroying something of type T includes, among other things, altering X so that it is no longer of type T.

On this definition, the concept of destroying takes the place of the concept of killing, and the relevant type T could be something like *Embodied Human Organism*: there is a strong moral presumption against altering me so that I am no longer an embodied human organism, since such alteration involves destroying an embodied human organism.

However, I am reluctant to adopt this definition, because it is not clear to me how "altering X so that it is no longer of type T" is really a legitimate instance of destroying something of type T. Take a teenager named Xavier. Now let Xavier turn 20. Xavier has been altered so that he is no longer a teenager—that is, he is no longer of the type *Teenager*. But I find it very odd to say that when he turns 20, a teenager has been destroyed. Or take a British politician named Xantippe, who happens to be a Tory. Now let Xantippe switch political parties. Xantippe has been altered so that she is no longer a Tory—that is, she is no longer of the type *Tory*. But again, I find it very odd to say that when she switches parties, a Tory has been destroyed. I am willing to grant that when Xavier turns 20, there is one less teenager in the world, and that when Xantippe switches parties, there is one less Tory in the world. And if that is all one means by "destroy", then the definition above might work. But in that case, I am reluctant to adopt the definition because it relies on an unconventional sense of "destroy".

Perhaps the best way of avoiding the tension between (a) and (b) is to stipulate that the concept of serious moral status includes a strong moral presumption against performing an action that either *makes* a human disembodied or *keeps* a human disembodied. Then serious moral status would cover normal acts of killing, which count as performing an action that makes a human disembodied. But it would also cover any cases in which a disembodied soul is prevented from re-entering its original body. Although the latter sort of act is not technically killing, it seems very similar to killing. Let the term "skilling" be synonymous with "performing an action that keeps a human disembodied." For a vivid example of skilling, imagine that you have a heart attack and become disembodied, floating above your body and viewing it from a somewhat remote distance, such as the ceiling of the room you are in when the heart attack happens. I rush onto the scene, but instead of performing cardio-pulmonary resuscitation (CPR) to make your disembodied soul return to your body, I instead use my ray-gun to instantly incinerate your body. I have not killed you—after all, you may still exist disembodied, floating near the ceiling of the room. But I have skilled you, since I have performed an action that kept you disembodied.

Once the concept of skilling is in place, I believe that the tension between (a) and (b) can be resolved by insisting that serious moral status should be understood, throughout the essay, as including a strong moral presumption against both killing and skilling. So, for example, the conclusion of the main argument should be understood to entail the following claim: "If something is human, there is a strong moral presumption against killing it when it can be killed, and a strong moral presumption against skilling it when it can be skilled." If it turns out that humans cannot survive the death of their biological organisms, then the second clause will be superfluous; but if it turns out that humans can survive the death of their biological organisms, then the second clause will be informative.[15]

Finally, we should note that the strong moral presumption against killing may or may not be applicable to non-biological entities, like advanced computers, which are functionally indistinguishable from biological entities. The reason is that it is only possible to kill something that is alive, and advanced computers are not alive. This might seem to be a significant disadvantage of my account, since on some views of the nature of the mind and mental states, advanced computers actually have a mental life much like the mental life of humans, and seem to experience emotions similar to the emotions humans experience when facing death. Remember the computer named Hal in the movie *2001: A Space Odyssey*, or remember the computer system from the movie *Short Circuit*, who pleaded with humans that they "No disassemble Johnny Five!", or remember the young Indian girl, who is actually a young computer program, in the movie *Matrix 3: Revolutions*, who gets "murdered" by the evil Agent Smith. However, I am not especially concerned to address this problem, and for two reasons. On the one hand, I am not convinced by computational

[15]Chris Tollefsen pointed out to me that the complexities of the last few paragraphs could have been avoided "by adopting a manner of argument common to Catholic pro-life philosophers," in which "killing is ending something's existence in its substance sortal—not just *any* 'type T'." These philosophers "are usually willing to bite the bullet and say, with Aquinas, that the disembodied soul is therefore not identical to—i.e., not the same person as—the embodied being whose life has been ended." Some of the reasons for preferring my view to this one are revealed in Chapter 6.

accounts of the mind, and hence remain agnostic about whether advanced computers would ever have the same capacities as a human being like Hillary Clinton. On the other hand, even assuming that advanced computers really did have minds, I am optimistic that a concept analogous to killing in the biological realm—perhaps "disassembly"—could be developed for such computers. The phrase "serious moral status" should be interpreted as including whatever concept analogous to killing applies to computers—if indeed there is such a concept. Likewise, if an advanced computer happened to have (or be) an immaterial soul—that is, if there really was a "ghost in the machine"—then a qualification about the difference between killing and skilling such a computer-soul would be in order.

4 What Are Typical Human Capacities?

The concept of a set of typical human capacities is a technical concept that relies upon the simpler concepts of a capacity and a hierarchy of capacities.

A capacity is the metaphysical ground or truth-maker of a true conditional statement about a thing. For example, the capacities of sub-atomic particles are what make statements describing what would happen to these particles in different circumstances true. Both ordinary language and the language of philosophy and science are suffused with statements about capacities. For example, you have the capacity to speak English, my pet rat has the capacity to experience pain, and salt has the capacity to dissolve in water. Such capacity-statements typically have a structure containing three elements: one which refers to the capacity itself, another which refers to the bearer of the capacity, and a third element which refers to what the capacity is a capacity for. Since the meaning of a given capacity statement is a function of these three elements, a word about each of them is in order.

First, there are various ways of characterizing the capacity itself. Modern discussions of capacities are sometimes carried out in terms of powers[16] or dispositions.[17] I shall understand the term "capacity" to be synonymous with such terms, as well as with terms like "capability", "ability", "potential", and "potentiality". For example, your capacity to speak English is the same thing as your ability or power or potential to speak English.[18] Although philosophers have sometimes used these various terms to make important distinctions among capacities, their ways of making such distinctions do not correspond to one another. I shall therefore use the previous terms

[16]See, for example, Harre and Madden (1975) and Molnar (2003).

[17]See, for example, the collection of key twentieth-century essays on dispositions in Tuomela (1978), Prior (1985), and Mumford (1998). A recent three-way debate on dispositions is found in Armstrong et al. (1996).

[18]There are different views about the nature of capacities. Some say they are ontologically primitive, and do not admit of further analysis. Others say they are ontologically derivative, because they logically supervene on "categorical" properties plus the laws of nature. Which of these views is more plausible will not affect the truth of the main argument.

synonymously, and shall mark distinctions among capacities by supplying further qualifying words like "active" or "passive".[19]

Second, ordinary language recognizes various ways of characterizing the bearers of capacities. The paradigm capacities are borne by individuals: for example, your capacity to think is something you have as an individual. However, not all capacities are borne by individuals: for example, the capacity of salt to dissolve in water is something the salt has as a roughly defined aggregate of stuff.

Third, there are different ways of referring to what a bearer of a capacity has a capacity for, or a capacity to do, or a capacity to be. For example, dispositional terms get paired up with terms referring to actions (e.g., "thinking" or "breaking"), properties (e.g. "hot" or "hungry"), and individuals (e.g. "this human being" or "this statue"). Dispositional terms also sometimes get paired up with phase sortals and substance sortals, which are count-nouns that serve as classificatory concepts for describing the world. As David Wiggins explains, the difference between phase sortals and substance sortals is the difference "between sortal concepts which present-tensedly apply to an individual *x* at every moment throughout *x*'s existence, e.g. *human being*, and those which do not, e.g. *boy*, or *cabinet minister*."[20]

With these three elements of capacity-statements before us, we are now in a position to notice two distinctions among capacities that are relevant to the main argument of this book: the distinction between active and passive capacities and the distinction between what might be called "identity-preserving" capacities and "compositional" capacities. Both of these distinctions can be traced back to Aristotle's discussion of potency. Aristotle wrote that all genuine potencies "are originative sources of some kind, and are called potencies in reference to one primary kind of potency, which is an originative source of change in another thing or in the thing itself *qua* other."[21] I shall label the capacities defined in terms of originative sources of change *identity-preserving* capacities, to emphasize the idea that their actualization preserves the identity of their bearer.

---

[19]As an example of how philosophers sometimes use these different terms to mark distinctions, Chris Tollefsen suggested that I make a distinction between dispositions—which (on his proposal) are "potentialities that can be possessed by all manner of things," and capacities, which (on his proposal) "seem to be dispositions of living things." One suggested reason this distinction "seems apt" is that "we tend to think of those dispositions, actualization of which is in some way or other under the control of the subject of the disposition, as being capacities. Not so dispositions which rely entirely on the activity of some other agent." So, for example, "the tendencies of human beings just qua bodies are not usually described as capacities": your tendency to fall towards the ground when dropped from a medium height is not (on this proposal) really a capacity of yours, but a mere disposition. However, I am reluctant to adopt Tollefsen's proposal. Both philosophical usage and common usage of the dispositional family of terms is highly variable. I find some who would want to call their tendency to lose their temper a "disposition" of theirs, and some who would want to call the tendency of a truck to carry a certain amount of weight a "capacity" of the truck. Therefore, I think it remains preferable to stick with my stipulation that I will be using all such dispositional terms as synonymous with "capacities", and that I will mark distinctions among capacities by adding extra terms.

[20]Wiggins (1967, p. 7, 1980, 2001).

[21]Aristotle (1941, 1046a8-1).

Aristotle thought that the primary sorts of potencies are "active" because their locus was "in the agent, e.g. heat and the art of building are present, one in that which can produce heat and the other in the man who can build."[22] Corresponding to these active potencies were "passive" ones, so-called because their locus was "in the thing acted on. . .[e.g.] that which is oily can be burnt, and that which yields in a particular way can be crushed."[23] Such potencies "of being acted on" were still genuine because they are "the originative source, in the very thing acted on, of its being passively changed by another thing or by itself *qua* other".[24]

Aristotle was also interested in the kind of potency some sort of matter has when considered in juxtaposition with something created out of the matter.[25] Such compositional potencies do not imply strict numerical identity between whatever is the actual and whatever is the potential. Rather, they imply the "is" of composition or constitution. Just as there is a difference between the "is" of identity and the "is" of constitution, there is a corresponding difference between what might be called the "is potentially" of identity and the "is potentially" of constitution.

So, then, we can sum up these two distinctions among capacities as follows. "Identity-preserving" capacities are those whose actualization preserves the numerical identity of the thing that bears the capacity, while "compositional" capacities do not imply such strict numerical identity. For example, my capacity to stand implies that I am numerically the same thing whether sitting or standing, but a lump of bronze's capacity to "become" a statue does not imply that the statue is numerically identical to the lump. "Active" capacities are powers to change things, whether in the world or in the bearer of the capacity, while "passive" capacities are powers for being changed by things in the world or in the bearer of the capacity.

The claim that humans have a set of typical human capacities should be understood in terms of identity-preserving capacities that include both active capacities (such as your capacity to walk) and passive capacities (such as your capacity to feel pain). The material bits and particles that gradually enter your lungs and bloodstream, and that eventually compose your brain tissues and legs, do not have the identity-preserving capacities to feel pain and walk. But you do have these identity-preserving capacities.

There is a further distinction among identity-preserving capacities that deserves to be noticed, which I shall call the distinction between general and specific (non-general, relational) capacities. Consider an example. You and I both have the capacity to think. The capacity to think is a general capacity. But your capacity to think your thoughts is a capacity specific to you, while my capacity to think my thoughts is a capacity specific to me. This would be true even if, contrary to fact, you and I had thought about the same sorts of things for every instant of our waking lives. Nobody else in the universe has the capacity to think my specific thoughts.

---

[22] Aristotle (1941, 1046a26-28).

[23] Aristotle (1941, 1046a22-26).

[24] Aristotle (1941, 1046a12-14).

[25] Aristotle (1941, 1048a37-b4).

Consider now a second example. You and I both have the capacity to remember our experiences. The capacity to remember is a general capacity. But your capacity to remember your experiences is a capacity specific to you, while my capacity to remember my experiences is a capacity specific to me. This would be true even if, contrary to fact, you and I had experienced the same sorts of things for every instant of our waking lives. Nobody else in the universe has the capacity to remember your specific experiences.

The claim that humans have a set of typical human capacities in common should be understood in terms of general capacities. Clearly, I do not have your specific capacities, and you do not have mine. Both specific capacities and general capacities are helpful for understanding personal identity through time, in the sense that they both help explain how I continue to exist when I undergo some change—for example, when I begin thinking about philosophy. But neither the general capacities nor the specific capacities help explain why I am not identical to you. The general capacity to think cannot explain this, since we have that in common. The specific capacities we both have, to think only our thoughts and not each other's thoughts, cannot explain this, since the fact that these capacities are specific already presupposes that I am not identical to you.

At the beginning of this section, I claimed that the concept of a set of typical human capacities is a technical concept that relies upon the simpler concepts of a capacity and a hierarchy of capacities. The last few paragraphs explained the concept of a capacity. It is now time to explain the concept of a hierarchy of capacities.

A hierarchy of capacities is a group of capacities with three interesting features. First, some members of the group are "lower-order" capacities. Second, some members of the group are "higher-order" capacities. Third, the higher-order capacities are just capacities to obtain the lower-order capacities. Examples from chemistry and biology illustrate this. Liquid water has the lower-order capacity to evaporate, while ice has the higher-order capacity to obtain this lower-order capacity. A mature oak tree has the lower-order capacity to support a tree house, while a sapling has the higher-order capacity to obtain this lower-order capacity.

When it comes to human organisms, ordinary language often presupposes these hierarchies even though it partially obscures them. For example, some would say that I do have the capacity to speak English, but I do not have the capacity to speak Chinese, because I can speak English right now if I needed to but it would take me a great deal of time and effort to learn to speak Chinese. We can capture what is going on here by saying that I have a lower-order capacity to speak English, but I do not have a lower-order capacity to speak Chinese. I do, however, have a higher-order capacity to speak Chinese. More precisely, I have a higher-order capacity to obtain the lower-order capacity to speak Chinese.

A number of contemporary philosophers have noticed this ambiguity in our talk about capacities and have sought to overcome it by employing certain hierarchical distinctions. For example, Michael Tooley has at least two discussions in which he notices such an ambiguity and attempts to eliminate it by introducing a distinction. In one discussion, he notes that there is "a certain imprecision in everyday talk about capacities" and suggests that this imprecision can be resolved by a distinction between capacities "in a narrower and stricter sense" and capacities "in a broader sense":

> Suppose that someone asks whether Mary is capable of running a six-minute mile. One person might say: 'Certainly, she's capable of running a six-minute mile, but she would have to get back into serious training for a few weeks.' What he is saying might equally well be put: 'No, she's not capable of running a six-minute mile at present, but she would be if she went back into serious training for a few weeks.' The moral is that when we speak about what someone is capable of, we may, as the first sentence illustrates, be talking, not about what they are presently capable of, but about what they could become capable of in a reasonably short period of time...Ordinary talk about capacities and capabilities is therefore somewhat imprecise. One must distinguish between capacities in a narrower and stricter sense, and capacities in a broader sense that includes potentialities for acquiring capacities in the narrower sense.[26]

In another discussion, however, Tooley makes a three-fold distinction between immediately exercisable capacities, blocked or suppressed capacities, and potentialities. The distinction between "immediately exercisable capacities, and blocked or suppressed capacities" is illustrated by his example of Mary, only this time her inability to run is due to alcohol:

> To attribute an immediately exercisable capacity to something is to make a statement about how the thing would be behaving, or what properties it would have, if it were now to be in certain circumstances, or in a certain condition...Thus, to say that Mary is now capable of running a five-minute mile is to say that if Mary were now to try to run a five-minute mile, were appropriately dressed, were on a track where there is not too much wind, and so on, then she would succeed in running a mile in five minutes. This is what I mean by an immediately exercisable capacity...Suppose, however, that Mary is drunk. In one sense, she is no longer capable of running a five-minute mile. Yet one might still attribute that capacity to her. In such a case, one might speak of blocked capacities—the idea being that all of the 'positive' factors required for the immediately exercisable capacity are present, but there are also negative factors that prevent the exercise of the capacity.[27]

Tooley then claims that "It is very important not to confuse the concept of a capacity—whether in the narrow sense that covers only immediately exercisable capacities, or in the broader sense that also includes blocked capacities—with the concept of a potentiality":

> To attribute a certain potentiality to an entity is to say at least that there is a change it could undergo, involving more than the mere elimination of factors blocking the exercise of a capacity, that would result in its having the property it now potentially has. It may also be to say that there are now factors within the entity itself that will, if not interfered with, cause it to undergo the relevant change.[28]

These two discussions of Tooley's are slightly confusing when put side by side, because he uses the labels "narrow" and "broad" differently in each discussion. In the first discussion, the "narrow" sense of a capacity is contrasted with a "broad"

---

[26]Tooley (1983, pp. 106–107).
[27]Tooley (1983, pp. 149–150).
[28]Tooley (1983, p. 150).

sense that includes potentialities, whereas in the second discussion, the "narrow" sense of a capacity only includes "immediate" capacities, and is contrasted with a "broad" sense that does not include potentialities, but only "blocked or suppressed" capacities. What this illustrates, I believe, is that although Tooley is correct in noticing that there is "a certain imprecision in everyday talk about capacities," it is also the case that there is a certain imprecision in everyday talk about potentialities. Indeed, as these two passages from Tooley demonstrate, there are certain states that we are strongly tempted to call a capacity in one breath and a potentiality in another. It would be nice if there were ways of eliminating ambiguities in our talk about capacities, without invoking the concept of a potentiality. Fortunately, there are such ways.

Eric Olson, like Tooley, notices an ambiguity in our talk about capacities, but he attempts to eliminate it by introducing a distinction between "first-order" capacities and "second-order" capacities:

> There are two different senses in which something might have the capacity to do something. In one sense, someone has the capacity to swim (for example) if she has learned how to swim and is not paralyzed, unconscious, or otherwise handicapped. If you put her in the water, she can swim. We might call this a "first-order capacity". But there is also a sense in which even someone who has not learned how to swim has the capacity to swim if she *could* learn how to do it. In this sense all human beings who aren't somehow physically disabled have the capacity to swim. Butterflies, on the other hand, do not have the capacity to swim. They simply aren't built for swimming. Someone has a "second-order capacity" to swim if she could acquire a first-order capacity to swim.[29]

If we accept Olson's suggestion, then what Tooley calls "immediate capacities" may be re-named "first-order capacities", what Tooley calls "blocked or suppressed capacities" may be named "second-order capacities", and what Tooley calls "potentialities" may also be named "second-order capacities". On Olson's picture, since both Mary-when-out-of-shape and Mary-when-drunk could acquire a first-order capacity to run a speedy mile, both Mary-when-out-of-shape and Mary-when-drunk would have a "second-order" capacity to run a speedy mile.

I believe that something like the sort of numerical approach Olson introduces is a helpful supplement to Tooley's approach, because it emphasizes the fact that Mary's "potentiality" to run a speedy mile is really just a certain sort of capacity to run a speedy mile, and that this capacity is different only in degree, and not different in kind, from Mary's capacity to run a speedy mile when she is drunk (or, for that matter, from Mary's capacity to run a speedy mile when she is sober). However, I believe that Olson's approach forces us to ignore some relevant differences between Mary-when-drunk and Mary-when-out-of-shape, since it simply labels both of them a "second-order" capacity. Part of what is needed, it seems, is a way of allowing for there to be more than two "orders" of capacities.

To get such a numerical approach that allows for more than two orders of capacities, we may look to the philosopher C. D. Broad, who sought to elucidate what he calls the "popular-scientific" view that many of us hold by employing, among

[29]Olson (1997, p. 86).

other things, the concept of a hierarchy of dispositions. The "popular-scientific" view itself is this:

> We ascribe to a thing a certain inner nature, and we hold that its history is determined jointly by its inner nature and its external circumstances. . .Thus a Thing is conceived as a store of powers or dispositions.[30]

Broad employed a number of distinctions among dispositions in order to clarify this view. He claims that things in nature have *orders* of dispositions:

> A bit of iron which has been put inside a helix in which an electric current circulates acquires the power to attract iron-filings. A bit of copper, placed in similar circumstances, does not. Under certain other circumstances, e.g., if it be sharply hit or heated to a certain temperature, the bit of iron will lose the magnetic property. If we call the magnetic property a "first-order disposition", the power to acquire this property when placed in a helix round which a current is circulating may be called a "second-order disposition" specific to iron. For it is a disposition to acquire the first-order disposition under certain circumstances, and it is common and peculiar to bits of iron. Similarly, the power to lose the magnetic property when heated or sharply hit will be a second-order disposition of iron.[31]

This leads to his general definition of a higher-order disposition:

> A disposition of the second order is, in general, a disposition to acquire or to lose, under assigned conditions, a disposition of the first order. In the same way we could define dispositions of the third or higher order.[32]

Finally, Broad believes his account applies to minds as well as matter:

> The power of learning to talk is a mental disposition of at least the second order, for it is a power to acquire a power of doing something. One of the peculiarities of minds in general, and of human minds in particular, is that they start with very few first-order powers, but rather with powers to acquire powers.[33]

Before applying Broad's distinctions to the ambiguity noticed by Tooley, it is important to pause and notice a question that Broad's example of the bit of iron leaves open. Let A be the first-order disposition of being magnetic, B be the second-order disposition of acquiring this first-order disposition, and C be the second-order disposition of losing this first-order disposition. Let t be a time before the bit of iron has been placed in the solenoid, t* be a time after the bit of iron has been placed in the solenoid, but before it has been heated, and t** be a time after the bit of iron has been placed in the solenoid, and after it has been heated. The picture Broad seems to be committing himself to is, at the very least, this: at t, the bit of iron has B but not A; at t*, the bit of iron has A and C; at t**, the bit of iron again has B but not A. This is represented in the following chart, with question marks representing the questions that Broad's example seems to leave open:

---

[30] Broad (1933, p. 265).

[31] Broad (1933, p. 266).

[32] Broad (1933, pp. 266–267).

[33] Broad (1933, p. 267).

	t	t*	t**
A: the first-order disposition of being magnetic	No	Yes	No
B: the second-order disposition of acquiring A	Yes	?	Yes
C: the second-order disposition of losing A	?	Yes	?

A more robust picture would put "Yes" where the question marks are, whereas a more austere picture would put "No" in these places. I think that Broad's own position eventually favors the more robust picture, but I need not show that here. What is important for the present discussion is that I favor the more robust picture, and will consequently formulate arguments below concerning humans assuming that the more robust picture is true. If the more austere principle turns out to be true, the arguments below could be reformulated without loss to accommodate it.

The main reason I favor the more robust picture is that I think the more austere picture relies upon the following principle, which I think is incorrect:

(1) Whenever an entity X has a disposition D at t to acquire (lose) a disposition D* at t*, then once X acquires (loses) D* at t*, X no longer has D at t*.

I think (1) is incorrect because it is simply a version of the following principle:

(2) Whenever an entity X has a disposition D at t to acquire (lose) a property P at t*, then once X acquires (loses) P at t*, X no longer has D at t*.

(2) seems false. In general, whenever a disposition is manifested, the disposition still continues to exist. For example, imagine David Hume (the philosopher) and David Robinson (the professional basketball player) are arguing over whether Robinson has the capacity, or disposition, to slam-dunk a basketball. "Show me the money," says Hume. "Prove you can dunk." Robinson then jumps up and slam-dunks a basketball. It seems that Robinson has the capacity, or disposition, to slam-dunk even when he is in the very act of slam-dunking. Otherwise, it would not be possible to *prove* that he has the disposition by *manifesting* it. But it is. Robinson had the disposition, at t, to slam-dunk at t*. When he actually slam-dunked at t*, he still had the disposition, at t*, to slam-dunk. Indeed, at t*, Robinson is showing off that very disposition. Hence principle (2) is false. But this basic line of argument can be recast even when we move to principle (1) and higher-order dispositions. For example, imagine Hume and Robinson are arguing over whether Robinson has the disposition to slam-dunk, but Robinson is currently recovering from an ankle injury. "I cannot dunk it right now, because I'm injured," says Robinson, "but give me six weeks, and I will be capable of dunking it." Robinson has a higher-order disposition, at t, to acquire the lower-order disposition 6 weeks later at t* to slam-dunk. When Robinson actually acquires the lower-order disposition at t* to slam-dunk, he still has the higher-order disposition, at t*, to acquire the lower-order disposition to slam-dunk. Indeed, at t*, the fact that Robinson has the lower-order disposition *proves* that he has the higher-order disposition.

In any event, with Broad's suggestion in hand, we might now be able to say that what Tooley calls "blocked or suppressed" capacities, and what he calls "potentialities," are really just different sorts of "higher-order" capacities, and that one of the differences between them (although certainly not the only one) is a difference in just how high their order is.

George Molnar recently argued for a numerically characterized hierarchy of capacities in a way similar to Broad. Molnar argued that one of the senses "in which we need a hierarchical taxonomy of properties" is to be found in the way dispositional properties are arranged:

> We can talk of a behavioral disposition of an object, for example *being magnetized,* as a first-order power. We can also talk of an object's capacity to acquire a first-order power, for example *being magnetizable,* as a second-order power. We can also call an object's aptness to lose a first-order power, for example metal's *tendency to fatigue*, a second-order power. . .For the sake of clarity, from now on I will refer to. . .powers to acquire (or to lose) a power as 'iterated powers'.[34]

Let me sum up this discussion of a hierarchy of capacities by revisiting the example where a mature oak tree has the lower-order capacity to support a tree house. It seems that if a sapling has a higher-order capacity to support a tree house, then so does an acorn. But the acorn's higher-order capacity to support a tree house seems to be higher in order than the sapling's. A more informative way of describing the difference between the acorn's and the sapling's capacity to support a tree house involves assigning numbers to the orders of capacities before explicitly relating these orders to one another. Let us call the mature oak tree's capacity to support a tree house a "first-order" capacity. Let us call the sapling's capacity to support a tree house a "second-order" capacity—recognizing, as we must, that numbering it "second" is somewhat arbitrary since there are any number of stages between the sapling and the mature oak. Finally, let us call the acorn's capacity to support a tree house a "third-order" capacity—again, recognizing that numbering it "third" is somewhat arbitrary since there are any number of stages between the acorn and the sapling. (Although the exact numbers we assign to the orders are somewhat arbitrary, what is not arbitrary is that we assign a higher number to the sapling's order than we assign to the mature oak tree's, and that we assign a higher number to the acorn's order than we assign to the sapling's.) We can now characterize the difference between the sapling and the acorn by saying that, while the sapling's second-order capacity is a capacity to obtain the mature oak tree's first-order capacity, the acorn's third-order capacity is a capacity to obtain the sapling's second-order capacity.

Given this understanding of a hierarchy of capacities, where every additional order of the hierarchy makes reference to the orders beneath it, I shall call the capacity where a given hierarchy starts a "first-order" capacity, but shall make a further important distinction between "first-order" capacities, on the one hand, and "immediate" or "immediately exercisable" capacities, on the other. For example,

---

[34] Molnar (2003, pp. 32–33).

what a first-order capacity to speak English amounts to is the neurological base for speaking English, which one gets from learning the language and which one retains as long as one remains a healthy and functional organism. What an immediate or immediately exercisable capacity to speak English amounts to, on the other hand, is a first-order capacity to speak English, the exercise of which is not impeded by some transient condition. When I am awake and free from any physical impediments blocking my ability to speak, I have both a first-order and an immediate capacity to speak English. But when I am asleep, or when my mouth is full of sand, I have a first-order but not an immediate capacity to speak English.

The concept of a hierarchy of capacities, and the distinction between first-order and immediate capacities, help explain the sense in which I still have a capacity even when I am temporarily "incapacitated" in various ways. For example, since I can think right now if I need to, I have an immediate capacity to think. If I am asleep, under anesthesia, or comatose, I lose the immediate capacity to think, but retain the first-order capacity to think. If I get certain kinds of reversible brain damage, I lose the first-order capacity to think, but retain a second-order capacity to think. If I get certain other kinds of reversible brain damage, I lose the second-order capacity to think, but retain a third-order capacity to think. And so on.

In general, then, an individual has a capacity, of some order or other, to do some activity whenever it has the ability to obtain the immediate capacity to do that activity. In some cases, obtaining an immediate capacity is relatively easy: for example, a sleeping or anaesthetized person eventually wakes up. In other cases, obtaining an immediate capacity is more difficult: for example, learning Chinese takes time and effort on the part of the learner and on the part of teachers. In still other cases, obtaining an immediate capacity requires extensive medical assistance: for example, recovering from a coma or brain damage requires the help of highly skilled professionals with sophisticated technology.

For a thing to have "a set of typical human capacities," it is enough for that thing to have such capacities either at some lower-order level or at some higher-order level. "Typical human capacities," in turn, are those capacities possessed at the first-order level by any normal adult human person. I shall not seek to define which capacities are included in "a set of typical human capacities" any more precisely than this. So, for example, I shall not attempt to show that the set of typical human capacities includes the capacity to experience pleasure and pain, or the capacity for self-consciousness, or the capacity for rationality, or some combination of these capacities.

5 What Are Persons?

Typical human capacities, I claimed a moment ago, are those capacities possessed at the first-order level by any normal adult human person. This immediately raises the question of what a person is. Since my answer to this question governs the way I approach The Adventure of Reagan's Brain, and since it governs the way I frame several arguments below, it is important to be clear up front about what my approach involves. I believe that the concept of a person should be given a purely descriptive sense, and that this descriptive sense should be assigned to it, not based on moral considerations, but on considerations emerging from our considered judgment of when it is proper to apply the use of personal names like "Ronald Reagan" and the use of personal pronouns such as "me". This section expands upon these claims.

The term "person" is used in a number of different ways both inside and outside philosophical discussions. One of the most important distinctions in the way "person" is used is what Joel Feinberg calls the distinction between normative (or moral) personhood on the one hand and descriptive (or commonsense) personhood on the other. Here is how Feinberg explains the normative use of "person":

> To be a person in the normative sense is to have rights, or rights and duties, or at least to be the sort of being who could have rights and duties without conceptual absurdity. . .when we attribute personhood in a purely normative way to any kind of being, we are attributing such moral qualities as rights or duties, but not (necessarily) any observable characteristics of any kind—for example, having flesh or blood, or belonging to a particular species.[35]

Examples of proposals for using "person" in this normative way include the following:

> "X is a person" just means "X has the right to life"
> "X is a person" just means "X ought to be treated as an end in itself"

On the other hand, here is how Feinberg explains the descriptive use of "person":

> There are certain characteristics that are fixed by a rather firm convention of our language such that the general term for any being who possesses them is "person.". . .I shall call the idea defined by these characteristics "the commonsense concept of personhood." When we use the word 'person' in this wholly descriptive way we are not attributing rights, duties, eligibility for rights and duties, or any other normative characteristics to the being so described. At most we are attributing characteristics that may be a *ground* for ascribing rights and duties.[36]

Examples of proposals for using "person" in this descriptive way include the following:

> "X is a person" just means "X has reason"
> "X is a person" just means "X has consciousness"

There is nothing incoherent about claiming that X is a person in some descriptive sense while denying that X is a person in some normative sense. But as Feinberg's last sentence indicates, there are many examples of substantive proposals for the relationship between X's descriptive personhood and X's normative personhood.

I believe that our concept of a person should not be governed by moral considerations, but rather should be guided by our use of personal pronouns (especially first-person pronouns) in situations where we try to imagine what can happen to us. This is because I can easily attempt to answer questions like "would that individual

[35] Feinberg (1980, p. 186).

[36] Feinberg (1980, p. 187).

be me?" while bracketing questions like "would that individual be a rights-holder, or an end-in-itself, or a possessor of such-and-such moral property?" Even an amoralist can entertain questions of the first sort. I therefore think of persons in purely descriptive terms.

It is instructive to compare my method for defining "person" with the method employed by Michael Tooley in *Abortion and Infanticide*. Tooley's methodological starting point for defining "person" is the desirability of obtaining a certain sort of result:

> It is. . .very important to have some term that refers only to. . .entities that have a right to life, and do so in virtue of their present properties, rather than in virtue of their potentialities. The term 'person' will here be understood in such a way that it applies to all and only entities of this sort.[37]

His preferred method for defining "person" in order to achieve this result runs as follows:

> Instead of defining a person as an entity that possesses a right to life, and that does so in virtue of its present properties, rather than in virtue of its potentialities, one can first determine what properties, other than potentialities, suffice to endow an entity with a right to life. Then one can define the term 'person' as applying to all and only those things that have at least one of the relevant properties.[38]

In attempting to justify this method for defining "person", Tooley recognizes the objection that "it is right both that the term "person", as ordinarily used, is generally a descriptive term, and that it is potentially misleading to define it instead in such a way that it is changed into an evaluative term." Nevertheless, he believes the method is justified because "the assignment of descriptive content to the term "person" is ordinarily guided by moral considerations."[39]

There are two implications of adopting Tooley's method for defining "person". First,

> When this approach is adopted, the term 'person' functions—as I believe it ordinarily does—as a purely descriptive term. The fact that something is a person does not by itself, therefore, imply any ethical conclusions. In particular, it does not follow from the fact that something is a person that it has a right to life.[40]

The second implication, however, is this:

> The fact that the descriptive content of the term 'person' has been selected with certain moral considerations in mind does mean, however, that if one's moral views are in fact correct, then it will be the case that an entity is a person if and only if it has a right to life, and does so in virtue of its present properties, rather than in virtue of its potentialities.[41]

---

[37] Tooley (1983, p. 34).

[38] Tooley (1983, p. 35).

[39] Tooley (1983, p. 35).

[40] Tooley (1983, pp. 35–36).

[41] Tooley (1983, p. 36).

My proposal for using "person" is different than Tooley's, and for two reasons. First, I believe that his justification for using "person" the way he does is inconclusive. He recommends applying the term to all and only entities that have a right to life in virtue of their present properties. Thus for any entity we imagine in our philosophical thought-experiments, the way to know whether that entity is a person or not is to begin by asking the question: "does this entity have a right to life in virtue of its present properties?" If the answer to this question is "yes", then the entity is a person, and we can go on to investigate what properties the entity has as a way of determining what property or properties are necessary and sufficient for being a person. If the answer to this question is "no", then the entity is not a person, and we can go on to investigate what properties the entity has as a way of determining what property or properties are *not* sufficient for being a person. He thinks this does justice to an alleged fact about our linguistic conventions, namely, that our assignment of descriptive content to "person" is ordinarily guided by moral considerations. But I believe that we can do more justice to our linguistic conventions by saying, with Feinberg, that the assignment of descriptive content to the term "person" is sometimes guided by wholly moral considerations and sometimes guided by wholly non-moral considerations. Once this is done, we can then be explicit about which sort of considerations are going to guide our assignment of descriptive content to the term.

The second reason why my approach for defining "person" is different than Tooley's emerges from a point I made a few paragraphs ago. When determining whether to use personal pronouns like "I" and "me", I need not, and often am not, guided by moral considerations. Even if I was a purely selfish individual, caring nothing at all for rights, duties, or anyone but myself, I could still entertain and answer questions like "is this decision going to bring me the things that I want in ten years?" and "would that individual, ten years from now, really be me?" while bracketing questions like "would that individual, ten years from now, be a dirty rotten scoundrel?" and "would that individual, ten years from now, have a right to life?"

6 What Are We?

Does an appeal to the use of personal pronouns automatically resolve philosophical debates about the nature of persons and personal identity? It would appear that the answer to this question is No. For ordinary language is sometimes ambiguous about what "we" are, and people can disagree about how to best resolve this ambiguity. Imagine philosopher-kings one day rule the world, and set up a division of their executive branch whose job it is to enforce the "proper" use of personal pronouns such as "I" and "me". The Personal Pronoun Police (PPP) get to decide on which of the following statements from ordinary language to ban or regulate:

(1) When I was born, I was a really ugly baby.
(2) After I die, I shall be buried in a cheap coffin.

(3) After I die, I shall experience wonderful things.
(4) If I lose my mind because of this brain disease, then please treat me tenderly.

One year, advocates of a certain sort of psychological account of personal identity run the PPP. They crack down on (1) and (4), replacing them with what they take to be more philosophically consistent paraphrases:

(1p) When the human organism with which I am currently associated was born, it was a really ugly baby.
(4p) If I lose my mind because of this brain disease, then please treat the resulting human organism tenderly.

Another year, substance dualists run the PPP, and so crack down on (2), replacing it with

(2s) After I die, my body shall be buried in a cheap coffin.

Yet another year, materialists run the PPP, and replace (3) with

(3m) After I die, nothing is left of me to experience wonderful things.

The imaginary case of the Personal Pronoun Police is designed to emphasize two points. First, as (1)–(4) illustrate, the use of personal pronouns in ordinary language is frequently ambiguous between different philosophical accounts of personal identity, and hence there may be no straightforward answer to the question "what account of personal identity does ordinary language favor?" Second, no matter who is running the PPP from one year to the next, and hence no matter how ordinary language gets revised for the sake of consistency, there is a difference between consistency and accuracy. For suppose substance dualism is true, but materialists are successfully running the PPP. Everybody on earth can use the word "I" in such a way that (2) and (3m) are true. But this would leave untouched the truth that, after death, there is *something* which continues to think, remember, experience pleasure and pain, and so on.

How is one to proceed, given this disagreement about persons, personal pronouns, and personal identity? My proposed strategy for overcoming this disagreement is made clear by a pair of examples. First, imagine a teenager named Adam who believes that

(A) A person cannot survive the transition from age 19 to 20.

You can imagine that this belief was formed any way you like. Perhaps Adam spontaneously formed this belief in a fit of misinterpretation and credulity on his 15th birthday when he overheard a trusted middle-aged neighbor make the offhand remark that "Well, a person can really *live* as a teenager, but once you turn twenty, well, then your life is pretty much over." However his belief in (A) was formed, Adam now has it, and has it quite firmly. It is a central strand in his web of beliefs. He is thoroughly convinced of its truth. Since Adam also believes that he is a person, he easily draws the conclusion that he cannot survive the transition from age 19 to 20.

Adam's friends try to persuade him to give up this belief; but nothing seems to work. They tell him that nobody else shares the belief; Adam replies that nobody else knows the sober truth of the matter, and that he must courageously bear the burden of that truth alone. Friends of his who are themselves past the age of 20 assure Adam that they experienced no annihilation on their 20th birthday; Adam replies that they are deluded, that the persons they seem to remember being are different persons than the persons they now are, and that the persons that they seem to remember being no longer exist. They ask Adam to imagine how much fun he himself will have celebrating his 21st birthday, and how much suffering he will experience the morning after the celebration if he drinks too much alcohol; Adam replies that he "cannot look forward to the pleasures of, and cannot dread the pains of, that remote individual, since that remote individual will not be me." Before midnight on his last day as a 19-year-old, Adam bids farewell to his friends, asking them to remember him fondly. After the clock strikes midnight, his friends cheer for him and jokingly ask him how it feels to have survived the transition. He makes the following reply: "Dear friends, I ask you, do not call me by the name of that good person; my name is Cain, and though I am thankful for what I inherited from that person you once called Adam, yet I am not he; the misfortune of his end was the good fortune of my beginning."

What can possibly be said to Adam that would dislodge his belief in (A)? "Snap out of it"? Adam's example is relevant because most of the positions taken by philosophers about personal identity through time are like Adam's in the following way: even if you disagree with the position, it is very difficult to argue people out of the position if they are already convinced of it. This is because people who are already convinced of the position will have available answers for every attempt to persuade them out of the position, just like Adam had an available answer for his friends.

Next, imagine a second example, of a teenager named Ben who believes that:

(B) A person cannot survive the period from age 20 to 30.

You can imagine that this belief was formed any way you like. Perhaps Ben spontaneously formed this belief in a fit of misinterpretation and credulity on his 15th birthday when he overheard a trusted middle-aged neighbor make the offhand remark that "Well, a person can really *live* as a teenager, and they can really *live* once they get past their twenties, but during those twenty-something years, you basically don't have a life." However his belief in (B) was formed, Ben now has it, and has it quite firmly. It is a central strand in his web of beliefs. He is thoroughly convinced of its truth. Since Ben also believes that he is a person, he easily draws the conclusion that he cannot survive the period from age 20 to 30.

Ben's friends try to persuade him to give up this belief, using all of the techniques that Adam's friends used when trying to persuade Adam; but nothing seems to work. However, there is at least one strategy available for persuading Ben that is not available for persuading Adam. Whether it is persuasive to Ben depends on what other beliefs Ben has. The strategy runs like this:

(1) Ben exists at 19.
(2) Ben cannot have temporal gaps.
(3) Ben exists at 31.

Therefore,

(4) Ben cannot cease to exist at 20.

If Ben accepts (1) through (3), he must accept (4). And if he continues to believe that he is a person, he will realize that (4) is inconsistent with (B). Therefore he must give up (B).

When Ben is confronted with this strategy, it is always an option for him to revise his other beliefs to keep (B) intact. If a person is committed to a certain view about personal identity through time, they always have the option of scrapping other beliefs that threaten that view. One option Ben has of replying to his friends is this:

(1) Ben exists at 19.
(2) Ben cannot have temporal gaps
(3*) Ben ceases to exist at 20.

Therefore,

(4*) Ben cannot exist at 31.

Alternatively, Ben could give up his beliefs about objects not having temporal gaps:

(1) Ben exists at 19.
(3*) Ben ceases to exist at 20.
(3) Ben exists at 31.

Therefore,

(2*) Ben can have temporal gaps.

I offer arguments throughout this book that rely upon the sort of strategy Ben's friends use. So this book is directed to those who believe that an entity cannot have temporal gaps. This book is also directed to those who believe certain things about when an entity exists at two distinct times. I am at present rather unsure of the best way to argue for these beliefs. But I think that the claim that an entity cannot have temporal gaps is plausible, and that my intuitions about when an entity exists at two distinct times are more plausible than the intuitions my rivals have about this.

Recall the example that began this chapter, The Adventure of Reagan's Brain. To the question of "what happens to Reagan in this adventure?", I answered, firstly, that Reagan survives the adventure, and secondly, that Reagan maintains his moral status throughout the adventure. I am not suggesting that those who deny the first part of my answer are inconsistent. I am just suggesting that their views are incorrect. The strategy for showing the incorrectness parallels the strategy of Ben's friends:

(1) Reagan exists before the brain disease does its damage.
(2) Reagan cannot have temporal gaps.
(3) Reagan exists after the injection-plus-scan-in.

Therefore,

(4) Reagan cannot cease to exist when the brain disease does its damage.

Just as Ben could escape the conclusion of the argument above in various ways, I recognize that opponents of the argument in Reagan's case can escape its conclusion in various ways. But I am confident that many people, both philosophers and non-philosophers alike, will be attracted to claims (1), (2), and (3) in the argument concerning Reagan, just as they are attracted to claims (1), (2), and (3) in the argument concerning Ben. Likewise, I am confident that many people, whether philosophically trained or not, will be sympathetic to the view that Reagan retains his moral status during the intermediate phase of his life, after the brain disease has done its damage, yet before the injection-plus-scan-in. The remaining chapters in this book are an attempt to build upon these ideas, to consider what these ideas imply about human organisms like Reagan and ourselves, and to defend these ideas against rival approaches.

Chapter 2
Anything You Can Do, I Can Do Also: Humans, Our Capacities, and the Powers We Share

In this chapter, I defend the first step of the main argument: if something is human, it has a set of typical human capacities. A more precise way of putting this step is as follows: there is some set, H, of capacities, such that for any individual X, if X is human, then X has H. I defend this step in the face of three main problem areas: the obvious diversity of capacities among normal humans, the nebulous sense in which undeveloped humans have capacities, and the apparent absence of certain capacities among abnormal humans.

1 How to Compare Capacities Between Individuals

It is important for my argument in this chapter and the next one to have a way of making comparisons between the higher-order capacities of different individuals. Fortunately, the concept of a hierarchy of capacities can be clarified by the concept of the "developmental distance" an individual must cross before it has a certain first-order capacity, and this latter concept—the concept of developmental distance—allows us to make such comparisons. Such developmental distance can be characterized either in terms of a certain number of temporal units or in terms of a certain number of developmental steps. The following two examples illustrate this.

First, assume that I acquired the first-order capacity to think on my 3rd birthday. Assume also that I had to form 10 new "nested" neural pathways—in which the formation of the second pathway requires the first pathway already to be formed, and the formation of the third pathway requires the second pathway already to be formed, and so on—in between my 2nd and 3rd birthdays en route to acquiring this first-order capacity to think. A characterization of the developmental distance I had to cross, on my 2nd birthday, before I had the first-order capacity to think, could take one of two forms. In terms of temporal units, the developmental distance I-on-my-2nd-birthday must cross before I have the first-order capacity to think is 1 year; in terms of developmental steps, the developmental distance I-on-my-2nd-birthday must cross before I have the first-order capacity to think is 10 pathways.

R. DiSilvestro, *Human Capacities and Moral Status*, Philosophy and Medicine 108, DOI 10.1007/978-90-481-8537-5_2, © Springer Science+Business Media B.V. 2010

Now consider a second example. Assume that I injured my brain on my 22nd birthday, and that I eventually re-acquired the first-order capacity to think on my 23rd birthday. Assume also that I had to form 10 new "nested" neural pathways in between my 22nd and 23rd birthdays en route to re-acquiring this first-order capacity to think. As before, a characterization of the developmental distance I had to cross, on my 22nd birthday, before I had the first-order capacity to think, could take one of two forms. In terms of temporal units, the developmental distance I-on-my-22nd-birthday must cross before I have the first-order capacity to think is 1 year; in terms of developmental steps, the developmental distance I-on-my-22nd-birthday must cross before I have the first-order capacity to think is 10 pathways.

The steps-based way of characterizing developmental distance is connected to the concept of a hierarchy of capacities. Roughly, a single developmental step corresponds to a single "order" of capacities in a given hierarchy of capacities. Consequently, in the pair of examples above, I have a 10th-order capacity to think on both my 2nd birthday and on my 22nd birthday. As I move one developmental step closer to having the first-order capacity to think, I thereby move one order lower in my hierarchy of capacities related to thinking. The final developmental step I take in acquiring the first-order capacity to think is the step from the second-order capacity to think to the first-order capacity to think.

Since there are different ways of individuating developmental steps, there are therefore different ways of individuating orders in a hierarchy of capacities. In the pair of examples above, instead of focusing on neural pathways, we could have focused on some smaller step, such as cell divisions, or some larger step, such as specific systems of neural pathways. Consequently, the number of orders in a given hierarchy of capacities is a function of just how fine-grained one wants the hierarchy to be. A hierarchy can have a handful of orders, or billions of orders, depending on how it is described. But this observation poses no problem at all to making informative comparisons between the higher-order capacities of different individuals. When comparing the higher-order capacities of two individuals, X and Y, we must be alert to how fine-grained the hierarchy of capacities is for X, and how fine-grained the hierarchy of capacities is for Y. When seeking to compare "apples to apples" and "oranges to oranges", the ideal case would be one in which X and Y are always exactly the same numbers of developmental steps from having a certain first-order capacity, no matter how fine-grained the hierarchy is. But other cases are also permissible. X and Y might have the same number of specific systems of neural pathways to build before having the first-order capacity to think, even though they have a different number of cell divisions to accomplish before having that first-order capacity.[1]

There is not a necessary connection between a given time-based way of calculating developmental distance and a given steps-based way of calculating developmental distance. Two examples illustrate this. First, it is possible for two

[1] More generally (if more pedantically), even if X and Y are the *same* number of developmental steps from having a certain first-order capacity, on a *less* fine-grained way of individuating orders in a hierarchy of capacities, they might be *different* numbers of developmental steps from having the first-order capacity, on a *more* fine-grained way of individuating those orders.

individuals, X and Y, to be the same developmental distance from having the first-order capacity to think, on a given time-based way of calculating developmental distance, yet be different developmental distances from having the first-order capacity to think, on a given steps-based way of calculating developmental distance. This would happen if X and Y both take 1 year to develop to the point where they have the first-order capacity to think, but X takes 10 steps during this 1 year and Y takes 20 steps during this 1 year. Second, and conversely, it is possible for X and Y to be different developmental distances from having the first-order capacity to think, on a given time-based way of calculating developmental distance, yet be the same developmental distance from having the first-order capacity to think, on a given steps-based way of calculating developmental distance. This would happen if X and Y both take 10 steps to develop to the point where they have the first-order capacity to think, but X takes 1 year to take these 10 steps while Y takes 2 years to take these 10 steps.

However, although there is not a necessary connection between a given time-based way of calculating developmental distance and a given steps-based way of calculating developmental distance, we are usually on safe ground in assuming time- and steps-based ways of calculating developmental distance to mirror one another. So, in the pair of examples above, we would have been on safe ground assuming that me-on-my-2nd-birthday and me-on-my-22nd-birthday were the same number of developmental steps away from having the first-order capacity to think, even if we did not know for sure the exact details of what those developmental steps were. In general, if X and Y are both human organisms, and if X and Y are the same developmental distance from having the first-order capacity to think, on some time-based way of calculating developmental distance, then it is usually safe to assume that X and Y are the same developmental distance from having the first-order capacity to think, on some steps-based way of calculating developmental distance.

Consequently, even if we did not know for sure the exact details of what the developmental steps were that I had to cross between my 2nd and 3rd birthdays on the one hand and my 22nd and 23rd birthdays on the other, and even if all we knew was the times it would take for me-on-my-2nd-birthday and me-on-my-22nd-birthday to acquire the first-order capacity to think, we would have been on safe ground assuming that me-on-my-2nd birthday and me-on-my-22nd birthday had equivalent "higher-order" capacities to think.

2 A Temporary Change Argument About What We Are

Now that the concept of developmental distance in place, I am in a position to defend the claim that if something is human, it has a set of typical human capacities. Recall the three main problem areas here. The first concerns the obvious diversity of human capacities among normal humans—such as you, me, Hillary Clinton, and Barack Obama. I think this problem can be solved without much difficulty by what I shall call "the normalizing approach." You and I, Clinton and Obama, are each what we might call normal adult human persons. In spite of our differences in first-order capacities, we normal adult human persons have a set of first-order capacities in common. I shall not try to give an exhaustive list of them. But the first-order capacity to think, the first-order capacity to communicate using language, the first-order capacity to pray, and the first-order capacity to understand humor would likely be among them. Possession of these and other first-order capacities is precisely what makes some human a "normal" adult human person. Let the set of capacities that all normal human adult persons have at the first-order level be called "the set of typical human capacities." I recognize that not all humans have the set of typical human capacities at the first-order level. But what I shall argue is that all humans do have the set of typical human capacities *at some order or other*.

You are human. You are a human organism. But you have thousands of properties right now besides the property of being human. For example, you have a certain shape right now, a certain mass right now, and a certain mental state (thinking) right now. Let us call all such properties your immediate properties. Among your immediate properties are your immediate capacities. For example, you have the immediate capacity to think right now, the immediate capacity to smile right now, and the immediate capacity to breathe right now. You have these and thousands of other immediate capacities right now. I would like to explore some of the ways in which you could lose, and then regain, some of these immediate capacities, as well as some of the ways in which you could lose, and then regain, some of your other immediate properties. In particular, I would like to focus on what sorts of capacities you hold on to if all of your immediate properties are lost, and then regained, besides the property of your humanity. I am going to argue that, even when we temporarily strip away your other immediate properties in this way, you still retain a certain set of typical human capacities at some order or other.

For expository purposes, I shall focus on just one of the typical human capacities: the capacity to think. Consider, then, the following three properties:

1. the property of having the capacity, at some order or other, to think
2. the property of having the immediate capacity to think
3. the property of being human

Now consider the following question: could you have the capacity, at some order or other, to think, even when you do not have the immediate capacity to think?[2] There are at least three approaches for arguing that the answer to this question is "yes". The first approach focuses on times of your life when you have not yet thought: for example, when you were a preconscious fetus in your mother's womb. Unfortunately, this approach is not persuasive to people who are ambivalent about the idea that you ever existed as a preconscious fetus in your mother's womb. For if you do not exist at a certain time, then you cannot have any capacities (such

[2]Some of what follows in the remaining parts of this chapter and the next overlaps with my "Capacities, Hierarchies, and the Moral Status of Normal Human Infants," *Journal of Value Inquiry* 43: 479–492 (December 2009).

as higher-order capacities) at that time. The second approach focuses on times of your life when you no longer have the immediate capacity to think: for example, when your consciousness becomes lost in the later stages of a terminal brain disease. Unfortunately, this approach is not persuasive to people who are ambivalent about the idea that you would continue to exist in the later stages of such diseases. For, once again, if you do not exist at a certain time, then you cannot have any (higher-order) capacities at that time.

A third approach focuses on times of your life when you lose the immediate capacity to think but eventually regain it: for example, when you are temporarily unconscious due to being asleep, anesthetized, or comatose. We can call such times "temporary changes" in your immediate capacity to think. It seems that you still possess the capacity, at some order or other, to think during these temporary changes. Fortunately, this approach is persuasive to almost everyone, since almost everyone thinks that you would continue to exist during these temporary changes. Those who do not think you exist during the times of these temporary changes would have to deny one of the following two ideas, each of which is very difficult to deny: (1) You exist both before the time of the "temporary change" and after the time of the "temporary change"; (2) It is not possible for an individual to have temporal gaps in its existence; or, put differently, for any individual X and any time t, it is not possible for X to exist both before t and after t unless X exists during t.

Let us call this third approach the temporary change argument. The temporary change argument can be formulated more precisely by considering what happens to you as you move through three times, t_1, t_2, and t_3. Although you are human and have the capacity, at some order or other, to think at each of these times, you have the immediate capacity to think only at t_1 and t_3:

	t_1	t_2	t_3
Do you have the capacity (at some order or other) to think?	Yes	Yes	Yes
Do you have the immediate capacity to think?	Yes	No	Yes
Are you human?	Yes	Yes	Yes

The temporary change argument can then be set out as follows:

(1) If there is no other feature of yours that we could base your possession of the capacity (at some order or other) to think on at t_2, then we should conclude that, as long as you are human at a given time, you have the capacity (at some order or other) to think at that time.

(2) There is no other feature of yours that we could base your possession of the capacity (at some order or other) to think on at t_2.

Therefore,

(3) As long as you are human at a given time, you have the capacity (at some order or other) to think at that time.

It will no doubt be objected that the second step is mistaken, since there are other available features of yours that we could base your possession of the capacity (at some order or other) to think on at t_2: for example, the property of having a functioning cerebral cortex.

But this objection is mistaken. To see why, simply replace the question "Do you have the immediate capacity to think?" in the above chart with the question "Do you have a functioning cerebral cortex?". Imagine that you receive an injury to your cerebral cortex, which causes that organ to cease functioning for a period of time before it resumes functioning. You retain the capacity (at some order or other) to think during this period of time, even though you do not have a functioning cerebral cortex. It may be a "passive" higher-order capacity rather than an "active" one, if the damage to the cortex requires a good deal of external assistance, such as surgery, to fix (recall Ronald Reagan's hypothetical adventure). But it is still a bona fide higher-order capacity.

Perhaps it will be objected that the relevant property is not having a *functioning* cerebral cortex, but simply having a cerebral cortex. But this objection is also unsatisfactory, since the chart and the argument can be revised even when "functioning cerebral cortex" is replaced with "cerebral cortex". Imagine again that you receive an injury to your cerebral cortex, which causes that organ to cease functioning for a period of time. However, in this case, surgeons remove the cerebral cortex from your body for a period of time in order to study and repair it. (Although scientists are currently not able to do this sort of thing, there is no reason in principle to suppose that it cannot be done.) When your damaged cerebral cortex is out of your body, it seems that you do not have a cerebral cortex at all. Yet you still have the higher-order capacity to think during this time.

One might object to this thought experiment on the grounds that you still have a cerebral cortex during the period of time in which it is out of your body. After all, it is your cerebral cortex, even if it is in a laboratory across the room, or across the world. But the thought experiment can be revised to accommodate this objection. Imagine that the surgeons, after removing the cerebral cortex from your body and studying it, decide that it is too damaged to repair. They decide to make you a new cerebral cortex from stem cells, or from whatever the latest techniques in regenerative medicine happen to be. The surgeons destroy your original cerebral cortex in an incinerator, go on a vacation for 2 weeks, and then begin manufacturing a new cerebral cortex for you when they return. Clearly, during the 2 weeks when your surgeons are on vacation, you do not have a cerebral cortex at all. Yet you still have the higher-order capacity to think during this time.

Some will object at this point that my use of the personal pronoun "you" in the temporary change argument is being used far too liberally. For the states that make for personal identity, this objection goes, might not be preserved when the cerebral cortex is damaged, and are certainly not preserved when the cerebral cortex is destroyed. Consequently, this objection concludes, it is false that "you" have the higher-order capacity to think when your cerebral cortex is destroyed. For "you" do not exist when your cerebral cortex is destroyed.

I disagree with this objection, partly for reasons that have already been discussed, and partly for reasons to be discussed later in this chapter. Imagine you knew, in advance, that the cerebral cortex damage was going to occur, and that you knew, in advance, the various scenarios that might unfold when the surgeons try to repair it—including the destruction-vacation-reconstitution scenario. It seems to me perfectly rational for you to look forward to being "made whole" again, to look forward to thanking your surgeons once the whole thing was over, and to anticipate telling them how much you appreciate the fact that they made sure you were taken good care of even when they were on vacation. I recognize that this approach disagrees with the intuition some philosophers have when considering thought-experiments involving upper brain transplants, since the intuition is that wherever your upper brain goes, there you also go. But I am not convinced by that intuition. The upper brain is just like upper liver: it is a mere part of you that can be damaged, destroyed, and re-constituted before being reinserted into you.

There are two noteworthy implications that follow directly from the claim that as long as you are human, you have the capacity (at some order or other) to think. First, it follows that as long as you are already human, you already have the capacity (at some order or other) to think, even before your very first moment of having the immediate capacity to think. Variations on the temporary change argument, such as the case of the doctors incinerating the cerebral cortex while they go on vacation for 2 weeks, make the point that a person ("you") can exist while it is not having any experiences at all, and indeed while it has no physical traces of having ever had any experiences at all. There is no reason to think that this point is any less true before your first experience than it is after your first experience. Second, it follows that as long as you are still human, you still have the capacity (at some order or other) to think even after your very last moment of having the immediate capacity to think.

These two implications correspond to the first two approaches, considered at the outset of this section, for arguing that you have the higher-order capacity to think even at times when you do not have the immediate capacity to think. Recall how the first approach emphasized the time of your fetal development, and the second approach emphasized the time of your of terminal psychological deterioration. As long as you are human during these times, you have the higher-order capacity to think during these times.

These two implications, of course, also lead us right into the second and third problem areas mentioned at the outset of this chapter: human organisms at the beginning stages of life, and "marginal" human organisms, at various stages of life. The final two sections of this chapter tackle these two problem areas.

Before going on to these problem areas, however, it is worth pausing a moment to re-emphasize the fact that I have focused on the capacity to think merely for expository purposes. The structure of the temporary change argument can be repeated, not just for the immediate capacity to think, but for any set of immediate capacities you have—indeed, any set of first-order capacities you have, such as the set of typical human capacities. In other words, a temporary change argument can show that, as long as you remain human at a given time, you have a set of typical human capacities, at some order or other, at that time. If we drop the qualifiers "at a given time" and "at that time", the temporary change argument seems to show that, as long as you remain human, you have a set of typical human capacities, at some order or other.

3 The Capacities of Undeveloped Human Organisms

The temporary change argument is a philosophical tool for testing and sharpening your beliefs about what it takes for you to persist through time. It is somewhat different than the more popular philosophical tools that are used for this task. Most philosophical approaches to your identity over time begin with the assumption that you exist right now, and then proceed straightaway into an examination of different thought experiments that focus on what sort of events could bring about the end of your existence, or on what sort of events could have sparked the beginning of your existence. The temporary change argument begins with the assumption that you exist—or at least appear to exist—both right now and at some later time, and then proceeds into an examination of different thought experiments that focus on what sort of events could happen in between now and this later time. Thus while most philosophical approaches focus on the beginning and end of our existence, the temporary change argument focuses upon the middle, and then works its ways out.

However, there are at least two clusters of questions that need to be carefully explored before this temporary change approach can be fully endorsed. First, focus again on the capacity to think. Just how far back in our biological life can we truly speak of ourselves possessing the higher-order capacity to think? Can we speak truthfully about having the higher-order capacity to think as infants, or fetuses, or embryos? Such questions will be taken up in more detail in this section. Second, just what sorts of changes can we permit to happen to us before we deny that we still have the higher-order capacity to think? Can we speak truthfully about having the higher-order capacity to think after terrible brain diseases have taken their toll, or in individuals with severe genetic deficiencies? Such questions will be taken up in more detail in the final section of this chapter.

The claim that human infants, fetuses, embryos, and zygotes have a higher-order capacity to think is perfectly unobjectionable to many people. After all, just as an acorn has a higher-order capacity to do the things a mature oak can do (such as support a tree house), so too a human infant (fetus, embryo, zygote) has a higher-order capacity to do the things a mature human can do (such as think).

But now consider a controversial question: why should someone believe that a human infant (fetus, embryo, zygote) actually has *any* higher-order capacity to think at all? Those asking this question could *agree* with the claim that you possess a higher-order capacity to think when you go through a temporary change, and that you possess a higher-order capacity to think even before your very first moment of possessing the immediate capacity to think. But they could *disagree* with the claim that you possessed this higher-order capacity when you were an infant, on the grounds that you *never were* an infant. Perhaps the commonsense idea that you once were a human infant is simply false. After all, this commonsense idea is flatly incompatible with certain accounts of what constitutes your identity over time. Those asking the controversial question might doubt that the infant from which you developed had your set of higher-order capacities because they doubt that the infant from which you developed was *you*.

For those who remain skeptical that human infants have a higher-order capacity to think, I offer the following argument. Some adults are the same, in terms of the order of their capacity to think, as some infants. Now, since you could *become* one of these adults, it follows that you could have *been* one of these infants. Indeed, since you could *become* one of these adults, there seems to be no good reason to deny that you *actually were* an infant at one time, and that you had the capacity (at some order or other) to think at that time.

A natural objection can be raised against the first step of this argument. Admittedly, this objection goes, it is not hard to think of cases involving adults who do not have the immediate capacity to think: for example, a sleeping adult must first wake up before she can truly be said to have this immediate capacity. But in our world, this objection continues, most adults who lack this immediate capacity need very little in order to regain it: for example, most adults when sleeping have a 1st-order, or at most a 2nd-order, capacity to think. Comparable things can be said if the adults without the immediate capacity to think are drunk, or sedated, or under anesthesia. Nevertheless, the objection goes, it is surely more difficult to think of cases involving an adult whose order of the capacity to think is even remotely comparable to, much less the same as, the order of the capacity to think possessed by a human infant. The objection concludes that the first step of the above argument—the claim that some adults are the same, in terms of the order of their capacity to think, as some infants—must be a long reach at best and simply mistaken at worst.

But this natural objection can be answered by focusing on cases of human adults whose present inability to think is more serious than the sort of "inability" to think brought on by sleep, alcohol, or anesthesia.

Consider the case of a normal human adult, named Albert, who suffers such a severe form of brain damage that he must go through a long period of rehabilitation in order to regain the immediate capacity to think. To fix ideas, let Albert be almost exactly like Thomas Nagel's example of "an intelligent person [who] receives a brain injury that reduces him to the mental condition of a contented infant...[for whom] happiness consists in a full stomach and a dry diaper."[3] The only difference between Albert and Nagel's example is that, while it is an open question whether Nagel's brain-damaged adult can ever outgrow his unfortunate condition, it is certain that Albert can. Albert can be fully rehabilitated over time. To grasp how much time it will take for Albert to recover, imagine that, at the time of his brain damage, Albert was the parent of an infant Benjamin. Albert's brain damage is so severe that it will take Albert the same amount of time to regain his first-order capacity to think as it takes Benjamin to obtain that first-order capacity for the first time. It seems that the

---

[3]Nagel (1979, p. 6).

order of Albert's capacity to think is as high as the order of Benjamin's capacity to think.

The concept of "developmental distance" introduced earlier in the chapter can be used to reinforce this example. The way Albert and Benjamin have been described, Albert and Benjamin are the same developmental distance, in terms of length of time, away from having the first-order capacity to think. And if we wish, we can simply *stipulate* that Albert and Benjamin are the same developmental distance, in terms of developmental steps, away from having the first-order capacity to think. But this means that Albert and Benjamin have the same order of the capacity to think, since each developmental step maps directly on to an order of the hierarchy of capacities leading up to the first-order capacity to think. The case of Albert therefore supports the claim that some human adults are the same, in terms of their order of the capacity to think, as some infants.

The major obstacle to this line of argument comes from so-called "psychological" accounts of personal identity over time, which claim that, for a person at one time to be identical with some entity at a second time, there must be certain sorts of causal links between the person's mental states at the first time and the mental states of the entity at the second time. The defender of the psychological account of personal identity will resist the comparison of Albert to Benjamin. Such a defender will insist that it is not possible for Albert to be reduced to the capacity-equivalent of Benjamin, since the changes involved in the relevant temporary change would undercut Albert's personal identity through time. According to the defender of the psychological account of personal identity, if Albert really has been reduced to the mental level of an infant, Albert no longer exists. If there would be a possibility of further mental development that would parallel the development of the mental life of an infant, that would not be the rehabilitation of Albert (according to the defender of the psychological account) but the development of a new and different person. Hence Albert's case cannot be used to support the claim that some human adults are the same, in terms of their order of the capacity to think, as some infants.

I will reply to this objection a few paragraphs from now, by arguing that variations on the temporary change case give us excellent reasons for rejecting such accounts of personal identity over time. If I am right about this, the result is that some adults really are the same, in terms of the order of their capacity to think, as some infants. Now, since you could *become* one of these adults, it follows that you could have *been* one of these infants. Indeed, since you could become infantilized and "incapacitated" just like Albert, there seems to be no good reason to deny that you *actually were* an infant at one time, and that you had the capacity (at some order or other) to think at that time.

It might be thought at this point that, although this strategy works for *some* human infants, it cannot work for *all* of them, and in particular it cannot work for abnormal infants whose abnormality consists of a brain defect that prevents the infant from developing the first-order capacity to think. Since this thought will be carefully evaluated, and eventually rejected, in the next section, there is no need to evaluate it here. Instead, the focus of the next few paragraphs is on a different question. Can the argument from the temporary change be extended to human organisms less developed than human infants?

Although there might be skeptical concerns that emerge out of psychological accounts of personal identity, the basic comparison at the heart of the temporary change argument would still be sound even if the word "infants" is replaced with the word "fetuses". The basic idea is that, as we imaginatively alter the temporary changes of the adult organisms, we eventually reach a point where the order of the capacity to think possessed by the organism in the middle of a temporary change is the same as the order of the capacity to think possessed by a human fetus. For example, just as a fetus only has a (say) 1,000th-order capacity to think, so too we can envisage an adult in the middle of a temporary change who only has a 1,000th-order capacity to think.

Consider the case of a normal human adult organism named Caleb that suffers such a severe form of brain damage that he must go through a long period of rehabilitation in order to regain the immediate capacity to think. Just as Albert could be fully rehabilitated over time, so too Caleb can be fully rehabilitated over time. To grasp how much time it will take for Caleb to recover, imagine that, at the time of his brain damage, Caleb was the parent of a fetus named Drake. Caleb's brain damage is so severe that it will take Caleb the same amount of time to regain his immediate capacity to think as it takes Drake to obtain that immediate capacity for the first time. It seems that the order of Caleb's capacity to think is as high as the order of Drake's capacity to think. The case of Caleb therefore supports the claim that there are some adults who are the same, in terms of the order of their capacity to think, as some fetuses. And the temporary change argument still goes through when amended to refer to human embryos or zygotes. All one needs to do is to construct cases involving a human adult organism Ebert and his embryonic offspring Frank, or a human adult organism Gilbert and his zygotic offspring Harry.

I would now like to consider several of the most powerful objections to the line of argument developed just developed. Although all of these objections rely on certain positions about what it takes for a person to persist through time, there are differences between the objections.

Perhaps the most clearly stated objection to the argument developed above, whether one is thinking about infants, fetuses, or what have you, is an objection that emerges from Chapter 5 and 6 of Michael Tooley's *Abortion and Infanticide*. In Chapter 5, Tooley argues that a *person* is an entity that possesses at least one of the relatively permanent, non-potential properties that make it intrinsically wrong to destroy that entity, and that do so independently of that entity's intrinsic value.[4] His eventual account of these properties has the important implication that an individual can be a human organism at a certain time without being a person at that time. Then, in Chapter 6, Tooley argues that:

> X is a *potential person* if and only if X has all, or almost all, of the properties of a positive sort that together would be causally sufficient to bring it about that X gives rise to a person, and there are no factors present within X that would block the causal process in question.[5]

[4]Tooley (1983, p. 87).

[5]Tooley (1983, p. 168).

The relevant objection to the argument of the present chapter is found towards the middle of Tooley's Chapter 6. He says that the only argument he has encountered in support of the claim that "the destruction of potential persons is intrinsically wrong", aside from arguments showing that there is a prima-facie obligation to actualize "possible" persons, is an argument that runs as follows:

> Consider a normal adult human being that is in a coma due to brain damage. Unless the term 'capacity' is used loosely, we cannot speak of such an individual as having capacities for self-consciousness, for rational thought, etc. It has only the potentiality of re-acquiring those capacities. Nevertheless, one certainly does not think that it is morally permissible to kill such an individual. If it is possible for the individual to recover, it is just as wrong to kill him as it is to kill a normal adult human being who is not in a coma. So potentialities *do* have moral weight. And if they count in this context, if their presence serves to make it seriously wrong to destroy something, why should they not count equally in the case of potential persons?[6]

This argument is quite similar to the argument developed above. Although this argument involves a coma while the argument developed above did not, this difference is not important for present purposes.

Tooley's case against this argument consists of two points. His first point is that if the mentioned potentialities were sufficient to make it wrong to kill the organism that is in a coma, then they must be understood as "passive" potentialities in cases where brain damage can only be repaired by others. But this is too broad, Tooley argues, since it entails that it is wrong to destroy living human skin cells, which, because of cloning technologies, have the relevant "passive" potentialities too. Tooley's concern here will be dealt with in a more specific way regarding cloning in Chapter 4. The basic idea is that allowing passive potentialities to bear moral weight is unproblematic as long as the distinction between identity and material constitution is upheld. The passive potentiality *a* has to become *b* in an identity-preserving way, and the passive potentiality *a*'s matter has to constitute *b*, are different sorts of potentialities. A living human adult in a profound coma has the former sort; a living human skin cell has only the latter sort.

Tooley's second point is that the mentioned potentialities are not sufficient to make it wrong to kill the organism that is in a coma. His reasoning here is worth quoting in full:

> For the injury may have deprogrammed the organism's brain, with the result that while *it* will shortly revive, and enjoy self-consciousness, etc., it will not have any of the memories, beliefs, attitudes, personality traits, and so on, of the person previously associated with that human organism. An organism may continue to possess certain general potentialities after the associated person has been destroyed. In addition to general potentialities, there must be states of certain sorts standing in appropriate causal relations to corresponding earlier states—the types of states being those upon which personal identity depends. Accordingly, it is simply not true that general potentialities suffice to make it wrong to destroy a human organism in a coma, even though the organism was once a person. And if general potentialities are not sufficient in this case, the present argument has failed to provide one with any

[6] Tooley (1983, p. 203).

grounds for thinking that general potentialities suffice to make it wrong to destroy organisms that are not yet persons.[7]

It is worth pausing for a moment to ponder the implications of what Tooley is claiming in this second point. Tooley had written at the outset of *Abortion and Infanticide* that "it is very difficult indeed to arrive at a defensible position on abortion unless one is prepared to come to terms with the difficult issue of the moral status of infanticide."[8] What the present passage illustrates is that it is difficult to arrive at a defensible position on either abortion or infanticide unless one is prepared to come to terms with the issue of the moral status of killing human organisms that undergo certain sorts of temporary changes. Tooley's reasoning here anticipates one of the main thrusts of Chapter 3: the serious moral status of certain adult human organisms who undergo specific types of temporary changes, and the serious moral status of certain undeveloped human organisms, stand or fall together.

Tooley argues that the serious moral status of both sorts of human organism fall together. The reason is that, in the case that he is discussing, the states that he thinks are the basis for personal identity have been completely destroyed, so that the person who existed previously no longer exists, even though the same human organism exists. Tooley would then be able to claim, when confronted by a "temporary change" argument, that the person ("you") no longer exists once the relevant change has occurred. The change in question, on his view, is not temporary but permanent, because it destroys or annihilates a person who can never come back.

However, Tooley's reasoning can be challenged by challenging his claim that "In addition to general potentialities, there must be states of certain sorts standing in appropriate causal relations to corresponding earlier states—the types of states being those upon which personal identity depends." Remember Ronald Reagan. Imagine that a normal adult human organism suffers an accident that causes brain damage so serious that the supposedly relevant causal links for personal identity through time are severed. However, imagine further that the organism eventually comes to possess the very same *types* of brain states as it possessed before: the organism's favorite symphony before the accident eventually becomes its favorite symphony after the recovery; the scent of chocolate which the organism found so exhilarating before the accident eventually becomes just as exhilarating after the recovery; the friendships, the religious commitments, the idiosyncrasies of belief and behavior all eventually become the same after the recovery as they were before the accident. The organism, as it were, lives the same bits of its life over again, but experiences those bits as though it were experiencing them for the very first time.

Now, in a case such as this, is the most natural interpretation that the original person—Reagan, let's say—ceases to exist, and that a new person, indistinguishable in every way from Reagan, has come to be associated with the same organism that Reagan was associated with? Or is the most natural interpretation that the *same* person—namely, Reagan—has recovered his original personality traits? It seems

[7]Tooley (1983, pp. 204–205).

[8]Tooley (1983, p. 2).

that the latter interpretation is preferable, and thus the causal links some think are necessary for personal identity through time turn out to be unnecessary. For the case was described in such a way that, even though it is precisely those causal links that have been severed, the same person nevertheless remains.

I believe that there are at least two additional reasons, which even someone sympathetic with Tooley's approach might be able to accept, for believing that it is the *same* person—namely, Reagan—that exists after the organism's recovery. Both of these reasons emerge from other passages in *Abortion and Infanticide*. First, when Tooley considers the question "does an individual have to have desires at a given time in order for it to be the case that some things are in its interest?"[9], his answer is "No" for the following reason:

> Suppose that there were a disease that completely eradicated all desires in normal adult human beings, but only for a time. If John has contracted the disease, and is now in a desireless state, but will return to normal, and will then enjoy a life that he deems worth living, one surely wants to say that John's continued existence is in his interest. So individuals can have interests at a given time without having any desires at that time.[10]

This first passage is relevant because it describes a case in which John undergoes what can only be described as a temporary change in all of his desires. John has a certain set of desires at t_1. John exists in a desireless state at t_2. John regains his desires at t_3.

Now, desires are among the traits that give personalities their distinctive shape and structure. John's desires are arguably just as central and important to John as his memories, or beliefs, or attitudes. Thus if John can have interests even while he is undergoing a temporary change with respect to his desires, then it seems like he could have interests even while he is undergoing a temporary change with respect to his memories, or beliefs, or attitudes. Suppose that there were a disease that completely eradicated all memories (and/or beliefs, and/or attitudes) in normal adult human beings, but only for a time. If John has contracted the disease, and is now in a memory-less (and/or belief-less, and/or attitude-less) state, but will return to normal, we surely want to say that John's continued existence is in his interest. But for something to be in John's interest at a given time, it must be the case that John exists at that time. Therefore, it seems that John can still exist even if he is now in a memory-less (and/or belief-less, and/or attitude-less) state.

The second relevant passage comes in a section of the book whose aim is to lay out a case against the idea that something's capacities determine whether or not it is a person. Part of this case consists of the claim that possession of certain capacities is not *necessary* to make something a person. Tooley considers two types of brain damage that make it "impossible for the organism to enjoy any consciousness at all": in the first type, "the damage might involve the complete destruction of those

---

[9]Tooley (1983, p. 117).

[10]Tooley (1983, p. 117).

structures that are the positive, constitutional basis of consciousness and rational awareness"[11]; in the second type,

> The damage might leave those structures intact, but damage other parts of the brain so that the structures in question are isolated, with the result that it is impossible for the capacities for consciousness, and for rational awareness, to be exercised as long as the damage goes unrepaired.[12]

Tooley claims that even in the first type of brain damage, it is "unacceptable" to conclude that "if one were to kill the organism, or allow it to die, one would not be guilty of having destroyed a person."[13] The "crucial point," he says, is that

> It might very well be possible to repair such damage, and that the result of doing so might be an organism that not only was capable of rational awareness, but that had the memories, beliefs, attitudes, personality traits, and so on, characteristic of the person who previously existed.[14]

Tooley then asks a question: "Would the resulting individual be identical with the individual who existed prior to the damage, or merely be a replica?" And the answer he gives is this:

> The view that he would be a replica does not seem plausible. A person who revives from a coma is not a replica of the person who existed previously. Yet a coma may very well involve brain damage that temporarily destroys the constitutional basis for rational awareness. Why should one distinguish between cases where the damage can be repaired by the organism itself, and cases where it cannot?[15]

The possibility Tooley is envisioning here is similar in certain respects to the case of Reagan presented above. And the interpretation Tooley gives to this possibility is very similar to the interpretation I have suggested above in the case of Reagan. Tooley's idea is that even in cases where brain damage destroys the constitutional basis for rational awareness, the resulting person is identical with the person who existed before the damage occurred. The idea I proposed with Reagan was that even in cases where brain damage destroys certain causal connections, the resulting person is identical with the person who existed before the damage occurred.

The remainder of this second relevant passage consists of the conclusion Tooley draws from the possibility he envisions, where brain damage destroys the constitutional basis for rational awareness:

> Once it is granted that the resulting person is identical with the person who existed previously, the argument can be put as follows. If something can be a person only if it possesses a capacity for rational awareness, then it follows that in destroying an organism that has suffered the sort of damage described above, one cannot be destroying any person. But if the damage is repairable, and if the result would be the revival of the person who previously existed, then the destruction of the organism, by making impossible any such revival,

[11] Tooley (1983, p. 153).
[12] Tooley (1983, p. 153).
[13] Tooley (1983, p. 153).
[14] Tooley (1983, p. 153).
[15] Tooley (1983, p. 153).

> thereby destroys the person in question. Hence it cannot be the case that there is a person only where there is a capacity for rational awareness.[16]

This remainder of the second passage contains an important ambiguity. Although Tooley does admit that the person after the recovery is the same as the person before the injury, it is not clear that he wishes to commit himself in this passage to the view that there is any person, much less the same person, which presently exists during the time between the injury and the recovery. Nothing Tooley has said commits him to the view that the person endures through this period of time, and there may be signs that he does not hold this view: for example, notice how he uses the phrase "the revival of the person who previously existed" rather than the phrase "the revival of the person who presently exists". Perhaps Tooley believes that the person does not persist through this period of time, but rather, as it were, vanishes when the injury occurs only to reappear when the recovery occurs. If so, then his view about destroying this person requires the assumption that someone can destroy a person at time t even if that person does not presently exist at t. But this is a rather difficult assumption to make. Imagine that we could revive (resurrect?) Socrates by waving a certain magic wand, and that this was the only way to revive (resurrect?) him. The assumption we are considering implies that, if we destroy the wand, it follows that, since we have made impossible the revival of Socrates, we have thereby destroyed Socrates. But this seems mistaken. After all, Socrates was already destroyed, by hemlock. Our destroying the wand does not destroy Socrates a second time. Since attributing this assumption to Tooley is not the most charitable way of reading him, the discussion to follow will proceed as though he does not hold that assumption, and that he does hold the view that the person persists through the period of time between the injury and the recovery.

If the person does indeed persist through the period of time between the injury and the recovery, and if this is the basis for Tooley's argument that the capacity for rational awareness is not necessary for being a person, then there is an important qualification that needs to be made about the conclusion of his argument. Although the argument does seem to show that what we have called the "immediate" capacity for rational awareness is not necessary to be a person, it seems to show this by relying on what we have called a higher-order capacity for rational awareness: the organism Tooley describes has undergone a temporary change in its immediate capacity for rational awareness, but it still has a higher-order capacity for rational awareness, as is shown by the possibility of recovery. Therefore, Tooley's argument in this second passage does not show that what we have called a higher-order capacity for rational awareness is not necessary to be a person. It only shows that what we have called an immediate capacity for rational awareness is not necessary to be a person.

When this second passage is considered alongside the first passage, it appears that one should not make a metaphysical distinction between cases where the brain damage to be repaired involves the constitutional basis for rational awareness, and

[16]Tooley (1983, pp. 153–154).

cases where the brain damage to be repaired involves the constitutional basis for specific desires of the person, as in the case of John. That is, just as persons can persist even though the constitutional basis for their rational awareness has been temporarily destroyed, so too persons can persist even though the constitutional basis for their desires has been temporarily destroyed.

But once it is admitted that persons can persist even when both the constitutional basis for their desires and the constitutional basis for their rational awareness have been temporarily destroyed, it seems hard to deny that persons can persist even when the constitutional basis for the causal links connecting earlier mental states to later mental states has been destroyed.

In short, then, I believe that these two passages from *Abortion and Infanticide* actually help to motivate the interpretation I have been urging in the case of Reagan. Reagan himself, and not a mere replica, exists even after the supposedly relevant causal links have been severed. Hence these causal links are not necessary for personal identity.

Once we realize that these causal links are not necessary for personal identity, in situations where the memories, beliefs, attitudes, personality traits, and so on after the recovery are *the same as* before the accident, we also realize that these causal links are not necessary for personal identity, even in situations where the memories, beliefs, attitudes, personality traits, and so on after the recovery are *different than* before the accident. It might be objected at this point that a significant similarity of personality traits are necessary for personal identity through time, even if causal links are not necessary. But this is not true. Imagine that it takes 10 years for Reagan to recover his personality traits. Surely Reagan himself exists during this recovery period: after all, it is *his* recovery. But during this recovery period, Reagan's organism does not have most of the personality traits it had before the accident. In short, Reagan still exists even when these personality traits do not. Therefore these personality traits are not necessary for Reagan's personal identity. Since Reagan can still exist *during* a recovery phase without these personality traits, why should we doubt that Reagan can still exist *after* a recovery phase without these personality traits?

I would now like to move on to briefly discuss a number of other objections to the argument I developed above. Some philosophers argue that, even if it makes sense to think of your identity stretching back to one of the later stages of your mother's pregnancy, when you were a fetus, it does *not* make sense to think of your identity stretching back farther than this, when even the slightest traces of a mental life are absent. Consider, for example, the following thought experiment from Jeff McMahan's book *The Ethics of Killing*. McMahan attempts to cast doubt on the idea that "one *was* a tiny cluster of cells" by asking us to consider "whether one could ever *become* such an entity":

> Imagine that in some of us the process of biological development were somehow reversed. Those to whom this happened would begin to grow younger, in biological terms. Eventually they would revert to being babies and thereafter would have to be placed in artificial wombs in order to survive. As their brains reverted to the infantile and fetal stages of their development, their mental lives would become increasingly rudimentary and would eventually disappear altogether when their brains ceased to be capable of supporting consciousness.

> Suppose now that one were to face this prospect. It is instructive to ask oneself when in this process of biological regression one would cease to exist.[17]

After presenting the example, McMahan claims that

> For my part, I find it impossible to believe that I would still be around when what we may neutrally designate as my organism had been reduced to a microscopic network of cells from which any possibility of consciousness had vanished.[18]

In reply, I believe that modifying McMahan's example so that it involves a temporary change can challenge McMahan's intuition. Imagine that, in addition to our biologically based cycles of falling asleep and falling awake, human organisms also had biologically based cycles of shrinking down, in a manner like McMahan suggests, and growing back up, in which the steps of shrinking are reversed. To keep other aspects of our existence constant in this modified example, imagine also that, just as there is always psychological similarity between oneself before and after a period of being asleep, so too there is always psychological similarity between oneself before and after a period of being shrunk. If this were how we lived out our biological lives, it would not only be possible to believe that we still existed during one of these shrunken moments, but it would be quite likely that we would be unable to shake this belief, just as we are unable to shake the belief that we still exist during one of our sleeping moments. And since this is so, it should not be impossible to believe that we exist during the *first* of such shrunken moments.

Some philosophers argue that, while it may make sense to think of your history stretching back to your embryonic stage, it makes no sense to think of your history stretching back prior to the formation of the "primitive streak" at approximately 14 days after fertilization. The reason the moment of primitive streak formation is thought to be so important is that it is considered to be the moment when three types of identity-undercutting possibilities are closed off: the possibility of fission (e.g. twinning), the possibility of fusion (e.g. the production of chimeras), and the possibility of separating out a totipotent cell from an "early human embryo" (in other words, from a morula or blastocyst; but this particular possibility extends only a couple days into the embryo's existence, since at 4–5 days the cells are no longer "totipotent" but "pluripotent").[19]

These three possibilities are sometimes invoked to undermine the claim that genetic similarity and/or causal continuity are adequate bases for organismic identity through time, or to undermine the claim that morulas and blastocysts are even bona fide individuals. But to see how these three possibilities are invoked to undermine the claim that, just as you were once an infant, so too you were once a morula or blastocyst, consider a trio of thought experiments from Peter Singer and Helga Kuhse. With the possibility of fission, the thought experiment is this:

---

[17] McMahan (2002, p. 29).

[18] McMahan (2002, p. 29).

[19] Thanks to Chris Tollefsen for helpful comments on improving this paragraph.

> A man and a woman have intercourse, fertilization takes place, and a genetically new zygote, let's call it Tom, is formed. Tom has a specific genetic identity—a genetic blueprint—that will be repeated in every cell once the first cell begins to split, first into two, then into four cells, and so on. On day 8, however, the group of cells which is Tom divides into two separate identical cell groups. These two separate cell groups continue to develop and, some nine months later, identical twins are born. Now, which one, if either of them, is Tom? There are no obvious grounds for thinking of one of the twins as Tom and the other as Not-Tom; the twinning process is quite symmetrical and both twins have the same genetic blueprint as the original Tom. But to suggest that both of them are Tom does, of course, conflict with numerical continuity: there was one zygote and now there are two babies.[20]

With the possibility of fusion, the thought experiment is this:

> A man and a woman have intercourse and fertilization takes place. But this time, two eggs are fertilized and two zygotes come into existence—Mary and Jane. The zygotes begin to divide, first into two, then into four cells, and so on. But, then, on day 6, the two embryos combine, forming what is known as a chimera, and continue to develop as a single organism, which will eventually become a baby. Now, who is the baby—Mary or Jane, both Mary and Jane, or somebody else—Nancy?[21]

With the possibility of separating out a totipotent cell from a morula or blastocyst, the thought experiment is this:

> It is now believed that early embryonic cells are totipotent; that is, that, contrary to the 'identity thesis', an early human embryo is not one particular individual, but rather has the potential to become one or more different individuals. Up to the 8-cell stage, each single embryonic cell is a distinct entity in the sense that there is no fusion between the individual cells; rather, the embryo is a loose collection of distinct cells, held together by the *zona pellucida*, the outer membrane of the egg. Animal studies on four-cell embryos indicate that each one of these cells has the potential to produce at least one fetus or baby.[22]

These three phenomena are sometimes invoked in discussions of the metaphysics of early human development, as a way of showing that *no* human babies can trace their histories back prior to the moment when the primitive streak is formed. But these phenomena do not show this, since even in cases where fission, fusion, and/or totipotent separation occur, the relevant phenomena occur prior to the formation of the primitive streak. It would seem that the most these phenomena could ever show is that *some* babies cannot trace their histories all the way back to a moment of fertilization, but must be content to trace their histories back to a moment of fission, or a moment of fusion, or a moment where a totipotent cell is separated out.

Nevertheless, many argue that these three phenomena show much more than this. Their basic argument is this:

(1) No human individuals can trace their histories back prior to the time when events like fission, fusion, and totipotent cell separation can occur.

---

[20] Kuhse and Singer (1994, p. 69).

[21] Kuhse and Singer (1994, p. 70).

[22] Kuhse and Singer (1994, p. 71).

(2) Such events can occur up until the time when the primitive streak is formed.

Therefore,

(3) No human individuals can trace their histories back prior to the time when the primitive streak is formed.

But this argument is defective: if we suspend belief for a moment about the second premise, we will quickly realize that the first premise is false. Even if it was possible for me, 10 years ago, to undergo the sort of "brain bisection" described by Thomas Nagel,[23] or the sort of "division" described by Derek Parfit,[24] this possibility of fission would not undercut the claim that I can trace my history back farther than 10 years.[25] The same thing is true with fusion. Even if it were possible for you and I, 10 years ago, to undergo the sort of "brain fusion" described by science fiction novels, this possibility of fusion would not undercut the claim that you and I can trace our own individual histories back farther than 10 years. And the same thing is true with the separation of a totipotent cell. Even if it was possible for me, 10 years ago, to create a new human individual by merely separating one of my body cells from the rest of my body, this possibility does not undercut the claim that I can trace my history back farther than 10 years. In short, when it comes to adult human beings, like you and I, the mere logical possibility of fission, fusion, or totipotent separation at some time t does not undercut the claim that the adults can trace their identity through time back before t. Why should it be any different with early embryonic human beings, like morulas and blastocysts?

At this point, some philosophers fall back upon the claim that morulas and blastocysts are not yet genuine individuals because of the three possibilities mentioned above. They can then argue that, since you can only trace your history back to other individuals, it follows that you cannot trace your history back to a morula or a blastocyst. The key claim that these philosophers fall back upon is typically argued for as follows:

(1) No human individuals can exist so long as events like fission, fusion, and totipotent cell separation can occur.

(2) Such events can occur up until the time when the primitive streak is formed.

Therefore,

---

[23] Nagel (1979, pp. 147–164).

[24] Parfit (1984, pp. 245–248).

[25] This reply goes back at least as far as when Michael Lockwood considered Mary Warnock's statement in a television interview that "before fourteen days the embryo hasn't yet decided how many people it is going to be." He writes: "It is, however, independently clear that she [Warnock] is not personally inclined to set much store by that fact, morally speaking. And rightly so, in my opinion: any philosopher who took *that* as grounds for denying that early human embryos are human beings would, it seems to me, have to deny that ordinary adult human organisms are human beings, if Sperry and others turned out to be correct in claiming that the effect of cutting the corpus callosum, which links the two hemispheres of the cerebral cortex, is to transform one human being into two." See Lockwood (1998, p. 190).

(3) No human individuals can exist prior to the time when the primitive streak is formed.

But this argument is defective for the same reason the other argument was defective: the first premise is false. Even if it is possible for me to undergo fission right now, this possibility of fission does not undercut the claim that I am an individual right now. Similar things can be said regarding fusion and totipotent cell separation. Since these phenomena do not undercut the individuality of human adults, why should it be any different for human morulas and blastocysts?

Let me sum up where this chapter has taken us so far. I have been defending the claim that if something is human, it has a set of typical human capacities. The first problem for this claim was the obvious diversity of capacities among normal human adults. I claimed that, despite our obvious diversity, there is a set of capacities that all normal human adults have in common at the first-order level, and I labeled these "the set of typical human capacities." The second problem is whether undeveloped humans have this set of typical human capacities. I argued that, even though they do not have this set at the first-order level, undeveloped humans do have this set of capacities at some level or other. It is now time to turn to the third problem area, which concerns the apparent absence of certain capacities among abnormal humans, such as human organisms that are disabled, diseased, or genetically deficient in some way.

4 The Capacities of Damaged and Disabled Human Organisms

The problem with the claim I am defending—that if something is human, it has a set of typical human capacities—is made most clear with atypical human organisms. In what sense does an irreversibly comatose human organism—not just temporarily comatose, but irreversibly—have the capacity to think? In what sense did the real Ronald Reagan—not the hypothetical one in The Adventure of Reagan's Brain, but the real one—have the capacity to think in the final stages of his biological life, after Alzheimer's disease had done its characteristic damage? In what sense does an anencephalic human fetus—not a normal fetus, but a fetus born without an upper brain—have the capacity to think?

Abnormal human organisms, such as those whose mental capacities are compromised by disease, disorder, or accident, are sometimes described as "marginal". But the term "marginal" has also been used in a different sense, to describe any human organism, whether normal or abnormal, that happens to be "at the margins of life"—for example, at the very beginning or very ending stages of its biological life. The Argument from Marginal Cases (AMC), which will be discussed in Chapter 5, uses the term "marginal" in both of these senses, with the result that its class of "marginal cases" includes, on the one hand, disabled and diseased human organisms, and, on the other hand, normal yet undeveloped human organisms.

It might be thought that my strategy for addressing the capacities of undeveloped human organisms does not address other marginal cases. For example, there are human organisms in the late stages of dementia or Alzheimer's disease, human organisms in "persistent" or even "permanent" vegetative states, and human organisms in "irreversible" comas. Such human organisms, it might be claimed, do not even have a higher-order capacity to think.

This sort of problem is discussed by a number of writers on the AMC. Recall how talk about higher-order capacities is a mere notational variant on talk about the potential to have an immediate capacity. A number of writers on the AMC claim that, while grounding the moral status of marginal cases on their potential to have a certain capacity may indeed deal adequately with one sub-class of the marginal cases, it does not deal adequately with other sub-classes of the marginal cases. For example, Lawrence Becker writes that

> There does not seem to be a morally relevant characteristic that distinguishes all humans from all other animals...The assertion that the difference lies in the *potential* to develop interests analogous to those of normal adult humans is also correctly dismissed. After all, it is easily shown that some humans—whom we nonetheless refuse to treat as animals—lack the relevant potential.[26]

Likewise, Daniel Dombrowski writes that "if the potential for a developed mental life is the criterion for moral respect, then there will still be marginal cases to lend support to the AMC."[27] A few pages later, he re-emphasizes the point that no matter what criterion one picks—even if the criterion incorporates "potential"—there will still be marginal cases left over:

> The most marginal of the marginal cases—those human beings who never have manifested any interests and never will do so, who apparently do not experience needs, affection, aversion, hopes, or fears—are in a situation quite different from that faced by a retarded human being or an infant.[28]

In order to make this problem as strong as possible, it is worth returning to the writings of C. D. Broad. As indicated earlier, Broad defends the view that "One of the peculiarities of minds in general, and of human minds in particular, is that they start with very few first-order powers, but rather with powers to acquire powers."[29] These powers to acquire powers are the same thing as what this essay has called higher-order capacities. Just as higher-order powers are powers to acquire powers, so too higher-order capacities are capacities to acquire capacities. But Broad immediately qualifies this idea as follows:

> In this connexion it is important to distinguish between two cases, viz., the reversible and the irreversible. In the first case a power can be gained and lost and gained again repeatedly by appropriate changes in the external circumstances. A bit of iron can be magnetised, and demagnetised, and remagnetised, repeatedly. In the second case the substance has not the power to regain a certain power which it has lost, or to lose a certain power which it has

---

[26] Becker (1983), quoted in Dombrowski (1997, p. 2).

[27] Dombrowski (1997, p. 21).

[28] Dombrowski (1997, p. 26).

[29] Broad (1933, p. 267).

> gained. If you injure a man's brain in certain ways, his mind will lose certain powers, and there is no known way of restoring these powers to his mind.[30]

Since Broad uses a certain sort of brain-damaged human organism as an example of an individual that "has not the power to regain a certain power which it has lost," and since higher-order powers just are higher-order capacities, then it seems Broad is committed to the idea that certain sorts of brain-damaged human organisms do not have even a higher-order capacity to think.

So, then, a higher-order capacity to think just is the potential to acquire the first-order capacity to think; but many writers on the AMC claim that some marginal cases do not have this potential. Likewise, a higher-order capacity to think just is the power to acquire the power to think; but even Broad claims that some brain-damaged humans do not have such higher-order powers. If Broad and many writers on the AMC are right, then there are marginal cases of human organisms that do not have even the higher-order capacity to think.

I believe this problem can be solved. If we attend carefully enough to the notion of a hierarchy of capacities, then every last one of the marginal cases—including even those Dombrowski calls "the most marginal of the marginal cases"—can be seen to possess the higher-order capacity to think. It is only the present state of technology that makes it difficult for us to recognize that such marginal cases possess the relevant higher-order capacity. But the technologies of the future—for example, neural therapies, neurosurgeries, reconstructive brain surgeries, and genetic therapies—will allow such marginal cases to overcome their cognitive deficiencies. This means that such marginal cases possessed the capacity (at some order or other) to think all along. They did not lack the capacity but the technology for realizing the capacity.

Recall Thomas Nagel's example of an adult who suffers a life-changing accident that reduces them to having the mental life of a newborn infant. That human organism, while not possessing the first-order capacity to think, nevertheless does possess a higher-order capacity to have this first-order capacity. This higher-order capacity, admittedly, would not be realized unless certain conditions were to be met—namely, the correction of the brain damage. But this should not cause us to be skeptical about the existence of this higher-order capacity. A human organism can have a capacity even if certain physical conditions block its realization.

The key for properly analyzing the problem Broad raises is to notice how he ends the last sentence of his discussion (quoted above) of the man with brain damage: "there is no known way of restoring these powers to his mind." Just because there is no *known* way of restoring certain powers to a human organism (or, as Broad puts it, to the mind of a human organism), that does not mean there is no way *at all* of restoring those powers. For example, a thousand years ago the known ways of restoring the powers of thinking to a human organism were much fewer than the known ways of restoring such powers today. There are certain sorts of injuries to the brain that we can reverse with today's technology, but that would have been

---

[30] Broad (1933, p. 267).

irreversible a thousand years ago. If such an injury had occurred a thousand years ago, it would have been a mistake to claim that the injured organism did not have the power to regain the power to think. The organism did still have the power to regain the power to think. What was lacking was merely the technology to permit this higher-order power to be realized.

But notice what this entails. A thousand years from now the known ways of restoring the powers of thinking to a human organism will be much greater than the known ways of restoring such powers today. There are certain sorts of injuries to the brain that we cannot reverse with today's technology, but will be able to reverse a thousand years from now. If such an injury occurs today, it is a mistake to claim that the injured organism does not have the power to regain the power to think. The organism does still have the power to regain the power to think. What is lacking is merely the technology to permit this higher-order power to be realized.

Hence there is the logical space to claim, even in the case of the sorts of brain injuries Broad is concerned with, that the loss of certain powers—certain immediate or first-order capacities—is only "irreversible" in the sense that we do not currently possess the technology to reverse it. But this sort of irreversibility is not very impressive from a metaphysical point of view. It tells us more about the state of our society's medicine than it does about the state of our patient's mind.

The situation does not change even when an accident or disease destroys the entire upper brain of a human organism, wiping out all first-order capacities for experiences. Stem-cell therapies of the future, we can imagine, will enable the organism to grow back the relevant brain parts, in a way similar to the way a starfish that loses one of its leg parts is able to grow back those relevant leg parts.

It might be thought that this way of dealing with marginal cases is acceptable in cases involving brain damage, but not acceptable in cases involving inherited genetic conditions that prevent the capacity to think from emerging. But this thought is mistaken because the gap between brain damage and genetic condition is a gap that can be bridged.

The argument bridging the gap between brain damage and genetic condition can be summarized in this paragraph and spelled out in the paragraphs that follow. One can imagine special cases that collapse the distinction between brain damage and a change in genetic condition. If the way of dealing with marginal cases outlined above is acceptable in cases involving brain damage, then it is also acceptable in these special cases. But if the way of dealing with marginal cases outlined above is acceptable in these special cases, then it is also acceptable for cases involving inherited genetic conditions. Therefore, if the way of dealing with marginal cases outlined above is acceptable in cases involving brain damage, then it is acceptable for cases involving inherited genetic conditions.

Begin by imagining a special case that collapses the distinction between brain damage and a change in a human organism's genetic condition. A clear sort of case is where the brain damage is brought about by an accident that changes certain genes that code for specific brain functions. For example, imagine that a scientist works with a machine that produces massive amounts of different kinds of radiation. Let us call two of these kinds of radiation A-rays and B-rays. A-rays produce a genetic mutation in every cell of the human body such that the cells are genetically equivalent to the cells of a human organism with a genetically based brain handicap. B-rays reverse the effect of A-rays. One day the scientist inadvertently steps in front of the machine while it is emitting A-rays, and all of her cells promptly mutate.

If the way of dealing with marginal cases outlined above is acceptable in cases involving brain damage, then it is also acceptable in these special cases. After all, there does not seem to be a metaphysically relevant difference between the brain damage in such a special case and the brain damage in a more traditional case—for example, where the brain damage is the result of an injury to the brain from an automobile accident. There are certain differences in the details of how the brain damage came about and about how the brain damage can be repaired. But the special cases and the traditional cases are similar in that technological assistance is needed to repair the damage. In the traditional cases, it takes some sophisticated technology involving brain surgery to fix the injury caused by the automobile accident; but in this special case, it just takes a flip of the switch to turn on the B-rays and fix the genetic condition caused by the A-rays. Therefore, after the administration of the A-rays and before the administration of the B-rays, it makes sense to say that the science worker still has the higher-order capacity to think. And this is so, in spite of the fact that after the administration of the A-rays and before the administration of the B-rays, our science worker does have a genetic condition that prevents the capacity to think from emerging. The basic insight of this analysis is this: the mere fact that a human organism possesses a genetic condition that prevents the capacity to think from emerging does not threaten the claim that the human organism possesses a higher-order capacity to think.

But there is no reason to think that this basic insight is limited to cases where the genetic condition is changed or caused by an accident. Imagine a second woman who is the same age as the science worker, but who has never worked in the science lab. This second woman has precisely the same genetic condition that A-rays can cause and that B-rays can fix. However, in the case of this second woman, the genetic condition is not a mutation caused by A-rays but is an inherited condition she has always possessed. Surely, if the science worker has the higher-order capacity to think, so does this second woman. After all, for both women, it just takes a flip of the switch to turn on the B-rays and fix the genetic condition. In short, since there is not a metaphysically relevant difference between a genetic condition that a human organism gets from an accident and a genetic condition that a human organism gets from its parents, it follows that there is not a metaphysically relevant difference between a genetic condition that a human organism gets from a change and a genetic condition that a human organism inherits. Whatever is true about the higher-order capacities in cases involving a changed genetic condition will also be true in cases involving an inherited genetic condition.

What emerges from the last three paragraphs is this: if the way of dealing with marginal cases outlined above is acceptable in cases involving brain damage, then it is acceptable for cases involving inherited genetic conditions. When a human organism has an inherited genetic condition that prevents the capacity to think from emerging, this does not automatically mean that the organism does not possess a higher-order capacity to think. Rather, it simply signals the fact that our present technologies have not advanced to the point where the human organism's higher-order capacity to think can be realized. The genetic therapies of the future—whether they consist of B-rays or something else—will allow such organisms to realize their higher-order capacity to think.

This solution allows for a thing's passive capacities to count as members of the set of that thing's higher-order capacities. There are two reasons for allowing a thing's passive capacities to count in this way. First, a thing's passive capacities are just as real as its active capacities. Although their conditions for realization are different, the passive capacities are metaphysically on all fours with the active capacities. Second, there is no reason for thinking this basic parity between passive capacities and active capacities changes when one moves to a different level in a hierarchy of capacities. Just as an active first-order capacity is metaphysically on all fours with a passive first-order capacity, so too an active second-order capacity is metaphysically on all fours with a passive second-order capacity. This pattern continues at each level in a hierarchy of capacities. The result is that, even if a brain-damaged (or genetically handicapped) human organism does not have the active higher-order capacity to think, this says nothing about whether she has the passive higher-order capacity to think. If she has a passive higher-order capacity to think, then she has a higher-order capacity to think.

There are various objections to my solution to the problem of abnormal humans. One family of objections is that, by allowing a thing's passive higher-order capacities to figure in its set of capacities, the result is an unwelcome proliferation of capacities in the world: in particular, some non-human animals, some parts of human organisms, and even some lumps of inanimate matter now appear to have a set of typical human capacities. I will put off discussing these objections until Chapter 4, where I discuss the capacities of inanimate lumps of matter and human parts, and Chapter 5, where I discuss the capacities of nonhuman animals. In the final pages of this chapter, I would like to discuss a different sort of objection.

One objection to my solution to the problem of abnormal humans—which I shall call a "Scrooge-like" objection—is that my solution is too generous in ascribing a capacity to an organism. In particular, the Scrooge-like objection goes, my solution is willing to ascribe a capacity to an organism even when a great deal of external assistance is needed for that alleged capacity to be realized. According to this objection, it is an abuse of language to ascribe a higher-order capacity to think to a brain-damaged (or genetically handicapped) human organism, merely on account of the possibility of future technologies that could correct this brain damage (genetic handicap).

I call this objection "Scrooge-like" because of its metaphysical miserliness. The Scrooge-like objection is premature, and for two reasons. First, as Elizabeth Prior has shown, we should be careful not to let our ordinary-language usage of dispositional terms be decisive in situations where that ordinary-language usage conflicts with more precise scientific usage of dispositional terms. Two related points from Prior are especially relevant here. The first point concerns the relationship between a disposition and its background conditions; the second point concerns the relationship between a disposition and its "initiating cause".

First, consider the relationship between a disposition and its background conditions. Prior uses the disposition of water-solubility to argue that we are caught between two conflicting intuitions whenever we attempt to decide in what case a disposition may be truly ascribed to an item:[31]

(1) "an intuition toward linking a dispositional predicate very firmly to a particular set of conditions"
(2) an intuition "to treat dispositional predicates as incomplete predicates"

If we follow (1), "we will ascribe the disposition of water-solubility to salt because salt dissolves in water at stp [standard temperature and pressure], but we will say that phosphorous is not water-soluble because it does *not* dissolve in water at stp." If we follow (2), "we will say salt is water-soluble at stp, phosphorous is not water-soluble at stp, but phosphorous is water-soluble at high temperatures and pressures." Prior then gives two reasons for preferring intuition (2) to intuition (1). Negatively, the only reason we have for preferring (1) is that "we tend to connect an ordinary-language dispositional concept more closely to one particular set of conditions than to any other," and even this tendency is open to counterexamples. Positively, we should prefer (2) because

> Those possessing the most precise dispositional concepts—the scientists—use dispositional predicates in this way. When a scientist talks about solubility it is always solubility in a particular solvent at a particular temperature and pressure. If he fails to indicate explicitly a solvent, temperature and pressure, it is assumed that he is talking about solubility in 100 ml of water at stp.

So, then, the first relevant point from Prior amounts to this: although ordinary language may point slightly against attributing a disposition to an item, on account of nonstandard background conditions, nevertheless, scientists still use disposition predicates as incomplete predicates, predicates whose completion consists precisely in the filling in of details like background conditions.

The second relevant point from Prior is this: nonstandard initiating causes are to be treated in exactly the same way as nonstandard background conditions are treated. One of her examples here is the disposition of fragility: "Suppose we have a piece of steel...[that] will not shatter if struck with a blow of medium force...[but] will shatter if struck a blow of immense force...The question...is whether this [second] shattering is a manifestation of fragility." She argues that the same negative and positive reasons for preferring intuition (2) to intuition (1) apply here: the only point in favor of (1) is ordinary language, but (2) has scientific practice in its favor.

These two points from Prior are relevant to the objection under consideration because the technological assistance needed to restore a brain-damaged human organism's powers of thinking is simply a nonstandard initiating cause (and the same thing goes for a genetically handicapped human organism). Since an item can still have a disposition when nonstandard initiating causes are required to manifest

[31] All quotations in this paragraph refer to Prior (1985, p. 7).

that disposition, and since dispositions and capacities are the same thing, it follows that an organism can still have a capacity when nonstandard initiating causes are required to realize that capacity. And of course, there is no reason to think that this situation is any different for a higher-order capacity than it is for a lower-order capacity. Prior's analysis thus provides an excellent first reason for thinking the Scrooge-like objection under consideration is premature.

The second reason the Scrooge-like objection is premature is going to take a few paragraphs to summarize. To begin with, even the activities of a normal, healthy human adult require external assistance of some kind. This is true, for example, for every act of her thinking: if certain external conditions (e.g. temperature and pressure) were not within a certain range, she would not be able to think the way she does. The same could be said if certain entities are not present in her immediate environment. If you take away her supply of oxygen, she will soon be unable to think. But this does not mean that she does not have the capacity to think. It merely means that she requires external assistance of some sort if she is going to realize the capacity to think.

But the assistance of oxygen is not importantly different from the assistance of technology, as the following two cases make clear. First, imagine two normal adult human organisms, A and B, who differ only in their spatial location—everything else about them, molecule for molecule, is the same. A is on the surface of the earth and is enjoying thinking about philosophy, but B is a thousand miles above the surface of earth and is enjoying very little right now. Do A and B have the same capacities or not? B will soon be unable to think because of the lack of oxygen in her spatial location. But surely the fact that B needs this external assistance does not threaten the claim that B has the capacity to think. The existence of B's capacity to think does not depend on B's location in space. Of course, the realization of this capacity may indeed depend on B's location in space. But the existence of this capacity does not.

This truth about location in space is paralleled by a similar truth about location in time. Imagine now two brain-damaged human organisms, A and B, who differ only in their temporal location—everything else about them, molecule for molecule, is the same. A is, like you, living during a time when technology allows her sort of brain damage to be repaired; B is, unlike you, living a thousand years ago, during a time when the technology which would allow for her sort of brain damage to be repaired has not yet been invented. Do A and B have the same capacities—in particular the same higher-order capacities—or not? B will not be able to think because of the lack of technology in her temporal location. But surely the fact that B needs this external assistance does not threaten the claim that B has the higher-order capacity to think. The existence of B's higher-order capacity to think does not depend on B's location in time. Of course, the realization of this higher-order capacity may indeed depend on B's location in time. But the existence of this higher-order capacity does not. And a similar analysis would apply to inherited conditions.

It might seem, at first glance, that this second reason for thinking the Scrooge-like objection mistaken is in tension with the first reason for thinking the Scrooge-like objection mistaken. For while this second reason claims that a capacity (disposition) can exist even when the conditions required for its actualization (manifestation) are not present, the first reason claims that a capacity (disposition) is an incomplete predicate which is not definable apart from the conditions required for its actualization (manifestation). The way of resolving this tension, it seems, is the idea that a capacity (disposition) can exist as long as there are *some* conditions that will allow for its realization (manifestation). This idea preserves the intuition that capacities exist even in the absence of the conditions required for their realization, while still capturing Prior's insight that dispositions are incomplete predicates.

Interestingly, however, Prior objects to this idea, because she is concerned it will compromise the usefulness of our dispositional terms. As she puts the point,

> Dispositional predicates are useful because they divide up the world into those items which possess a particular disposition D and those items which do not...[but] would lose this utility if our criterion for ascribing disposition D to an item were simply that item would manifest that disposition under *some* set of conditions.[32]

I believe Prior's concern here is misplaced. Although dispositional predicates formulated using specific conditions may be *more* useful than dispositional predicates formulated using nonspecific conditions, this does not mean that the latter are useless. Indeed, the reasons Prior gives for thinking the former useful are also reasons for thinking the latter useful. She claims if we use dispositional predicates as incomplete predicates,

> These predicates will retain their utility. For in most cases the predicate 'has disposition *D* (under conditions C)' may be truly ascribed of some objects but *not* of others. Thus the predicate 'has disposition *D* (under conditions C)' helps us to divide up the world.[33]

What Prior seems not to notice is that the very same thing can be said about dispositional predicates when the parenthetical qualifier "under conditions C" is replaced with "under *some* set of conditions." In most cases the predicate "has disposition *D* (under some set of conditions)" may be truly ascribed of some objects but *not* of others. Take solubility. Although Prior recognizes that "virtually any solid will dissolve under some set of conditions,"[34] this still does not mean the dispositional predicate "soluble" would be useless when defined as "has disposition to dissolve (under some set of conditions)". For even on this definition, not everything in the world is soluble. For example, gases, liquids, fundamental particles, numbers, sets, and mental states are not soluble on this definition. Consequently, the predicate "has disposition to dissolve (under some set of conditions)" helps us to divide up the world.

In summary, then, the Scrooge-like objection claims that it is an abuse of language to ascribe a higher-order capacity to think to a brain-damaged (or genetically handicapped) human organism, merely on account of the possibility of future technologies that could correct this brain damage (genetic handicap). But there

---

[32] Prior (1985, p. 6).

[33] Prior (1985, p. 8).

[34] Prior (1985, p. 6).

are two reasons why this objection is premature: first, following Prior, a higher-order capacity to think can still exist even though non-standard initiating causes (like technological assistance) are needed before that capacity is realized; second, the assistance provided by such technology is not importantly different than the assistance provided by standard sorts of assistance, such as breathable oxygen.

It is in the context of marginal cases generally that it becomes most obvious that I have allowed for a thing's "passive" capacities to count as members of the set of that thing's higher-order capacities. And it is in the context of marginal cases with inherited genetic conditions that this allowance about passive capacities results in a position distinct from the pro-life Roman Catholic position. According to some leading Roman Catholic thinkers on this subject, such as John Finnis, Robert P. George, and Germain Grisez, it is the "active" higher-order capacity for a certain sort of thinking that makes an entity human: so if a human mother has a fetus growing inside her which does not have that segment of the human genome which encodes the molecular instructions for building the upper parts of the human brain, that fetus is not human. According to me, such a fetus is human because it has the same basic genotype as you or I, and it has a higher-order capacity to think because it has the ability to obtain the immediate capacity to think. This genetically deficient fetus has a passive higher-order capacity to think. But passive capacities count.

It is time to sum up the discussion of this chapter. Attending to the notion of a hierarchy of capacities explains why humans who undergo certain types of temporary changes—such as sleeping humans, anesthetized humans, and temporarily comatose humans—still have a set of typical human capacities. This notion also explains why another group of human organisms—such as human infants, fetuses, and zygoes—have a set of typical human capacities, even before they have exercised any capacities. However, it might be objected that attending to the notion of a hierarchy of capacities is not relevant to another sub-class of human organisms—for example, the irreversibly comatose, those in a persistent vegetative state, or those who are congenitally mentally deficient because of a genetic condition. Nevertheless, by attending more carefully to the notion of a hierarchy of capacities, all human organisms can be seen to have the higher-order capacities in question, even if not all human organisms realize those higher-order capacities.

If the argument of this chapter is correct, then if something is human, it has a set of typical human capacities. But why should this matter morally? What relevance do capacities—and the various metaphysical distinctions among capacities—have to the issue of serious moral status? After all, oak trees, saplings, and acorns may have a set of typical "oaken" capacities, at some order or other—and, if the argument of this chapter is correct, they may have such capacities even after being damaged by bad weather and termites—but this does not lead us to view these members of the plant kingdom as having serious moral status. So why should we think that the capacities of human organisms are any different? What is the connection between typical human capacities and serious moral status? The next chapter discusses this issue.

Chapter 3
The Only Game in Town: Why Capacities Must Matter Morally

In this chapter, I defend the second step of the main argument: if something has a set of typical human capacities, it has serious moral status. The major line of reasoning in this chapter is that there are cases of human organisms that are in such a state that the only satisfactory basis for their serious moral status is their set of typical human capacities. Since most of us—philosophers and non-philosophers alike—do believe that these human organisms have serious moral status, and since most of us do base this serious moral status on something, it must be the set of typical human capacities that we base it on. Finally, I spend the last two sections of the chapter explaining why the second step of the main argument can be arrived at in different ways, by adopting either the moral framework of John Rawls' original position or the moral framework of Martha Nussbaum's "capabilities approach".

1 A Temporary Change Argument About What Matters Morally

There are cases of human organisms that are in such a state that the best available basis for their serious moral status is their set of typical human capacities. Since most of us—philosophers and non-philosophers alike—do believe that these human organisms have serious moral status, and since most of us do base this serious moral status on something, it must be the set of typical human capacities that we base it on.

The human organisms I have in mind here are not human infants. Of course, if I wanted to, I could adopt a strategy that begins with the intuition that normal human infants have serious moral status. I could then argue that, since this serious moral status must be based upon *something*, and since the only plausible thing that it could be based upon is the capacities of the infants, it follows that human infants possess a set of capacities that generates serious moral status. An advantage of this strategy is that it appeals to an intuition that is widely shared by philosophers. But

R. DiSilvestro, *Human Capacities and Moral Status*, Philosophy and Medicine 108,
DOI 10.1007/978-90-481-8537-5_3, © Springer Science+Business Media B.V. 2010

a disadvantage of this strategy is that it cuts no philosophical ice with those who do not share the intuition.[1]

The human organisms I have in mind here are those who undergo various sorts of "temporary changes" like the temporary changes discussed in Chapter 2. Whereas Chapter 2 introduced what might be called a "metaphysical" temporary change argument, I would now like to defend what might be called a "moral" temporary change argument. While the metaphysical temporary change argument is supposed to show that being human is sufficient for having certain capacities, the moral temporary change argument is supposed to show that having these capacities is sufficient for having serious moral status.

This methodology, of course, immediately raises the following question: why not just focus on the property of being human, and avoid all these elaborate detours into the realm of capacities? After all, let us assume that temporary changes with humans lead us to recognizing the existence of two properties that might be sufficient for serious moral status. The first property, Q, is the property of having a set of typical human capacities. The second property, R, is the property of being human, which, as Chapter 1 argued, amounts to having the same basic genotype as you and I have and chimpanzees do not. Why prefer Q over R?

An answer to this question will emerge if we consider cases of what might be called "shumans" who undergo temporary changes. Shumans, let us say, are phenotypically indistinguishable from humans, but they have ZNA instead of DNA in their cells, so they have a different basic genotype than you and I. Shumans have all the same sorts of immediate capacities, and all the same sorts of higher-order capacities, as humans. Let us say that temporary changes with shumans lead us to recognizing the existence of two properties that might be sufficient for serious moral status. The first property is Q (the property of having a set of typical human capacities). The second property, S, is the property of being shuman, which amounts to having the same basic genotype as typical shumans with ZNA. Now, then, if S was sufficient for serious moral status, this would allow us to explain why shumans have serious moral status, but it would not help us to explain why humans have serious moral status. Conversely, if R was sufficient for serious moral status, this would help us to explain why humans have serious moral status, but it would not help us to explain why shumans have serious moral status. However, if Q were sufficient for serious moral status, this would help us to explain why both humans and shumans have serious moral status during their temporary changes. The fact that Q would have this

[1] Michael Tooley argued, in Chapter 10 of *Abortion and Infanticide*, that the intuition should be set aside for at least three reasons. First, appeals to moral intuitions are plausible only if the appeal is to principles that are (what Tooley calls) basic moral principles (or derivable from basic moral principles), but the principle appealed to by this intuition is not. Second, the intuition in question is not unanimously shared in our own society and was not generally shared by earlier societies. Third, the intuitions of people over the past 2000 years have been heavily shaped by the religions of Judaism and (especially) Christianity, so that a person can reasonably rely upon such intuitions only if he takes the teachings of one of these religions to be true (Tooley, 1983).

explanatory power is what makes it the best explanation for why both humans and shumans have serious moral status during their temporary changes. Q is preferable to R or S because Q explains more.

So, then, what is this "moral" temporary change argument? How does it work? Once again, for expository purposes I shall focus on just one of the typical human capacities—the capacity to think.[2] Consider the following question: do you have serious moral status when you do not have the immediate capacity to think? There are three approaches for arguing that the answer to this question is "yes". One approach focuses on times of your life when you have not yet attained the immediate capacity to think: for example, when you were an infant. Unfortunately, this approach is not persuasive to people who are ambivalent or undecided about whether human infants have serious moral status. A second approach focuses on times of your life when you no longer have the immediate capacity to think: for example, when you become psychologically incapacitated in the later stages of a disease like Alzheimer's. Unfortunately, this approach is not persuasive to people who are ambivalent or undecided about whether those who are in the later stages of such diseases have serious moral status. A third approach focuses on times of your life when you go through what Chapter 2 called "temporary changes" in your immediate capacity to think: for example, when you are temporarily unconscious due to being asleep, anesthetized, or comatose. Unfortunately, this third approach is sometimes left imprecise and undeveloped by its adherents. As a result, this third approach is often simply ignored or brushed aside by its detractors. It is often viewed as a conceptual wrinkle that can be easily ironed out while still preserving the central importance of the immediate capacity to think.

This third approach, which focuses on temporary changes, is actually far more promising than its detractors realize. Consider again times of your life when you lose the *immediate* capacity to think but eventually regain it: for example, when you are temporarily unconscious due to being asleep, anesthetized, or comatose. It seems that you would continue to exist during these temporary changes, that you would still possess the capacity *at some order or other* to think during these temporary changes, and that you would still have serious moral status during these temporary changes:

	t_1	t_2	t_3
Do you have serious moral status?	Yes	Yes	Yes
Do you have the immediate capacity to think?	Yes	No	Yes
Do you have the capacity (at some order or other) to think?	Yes	Yes	Yes

[2]But since the problem discussed in this section is a general problem, the discussion could be formulated in terms of any single capacity (e.g. the capacity to experience pleasure and pain) or any set of capacities (e.g. the set of capacities possessed by any normal adult human person).

The moral temporary change argument can now be set out as follows:

(1*) If there is no other feature of yours that we could base your possession of serious moral status on at t_2, then we should conclude that, as long as you have the capacity (at some order or other) to think at a given time, you have serious moral status at that time.
(2*) There is no other feature of yours that we could base your possession of serious moral status on at t_2.

Therefore,

(3*) As long as you have the capacity (at some order or other) to think at a given time, you have serious moral status at that time.

This "moral" temporary change argument has exactly the same form as the "metaphysical" temporary change argument of Chapter 2. And as before, it will no doubt be objected that the second step is mistaken, since there are other available features of yours that we could base your possession of serious moral status on at t_2. I have already explained why the property of being human is unsatisfactory for this purpose. Consider then four other available features of yours, besides the feature of having a capacity (at some order or other) to think, that might be thought satisfactory for generating serious moral status: (A) the property of having a certain sort of past; (B) the property of having a certain sort of future; (C) the property of having the first-order capacity to think; (D) the property of being an "actual, continuing subject of experiences." In the next four sections of this chapter, I shall be arguing that none of these properties, whether considered jointly or together, are satisfactory as properties for basing your possession of serious moral status on.

2 Moral Status and the Past

One strategy for replying to the temporary change argument is to ground your serious moral status at t_2 in the fact that certain things were true of you in your actual past: for example, you actually possessed the immediate capacity to think at times like $t_{1.}$ According to this first strategy, serious moral status only exists in the case of the temporary change in virtue of the fact that the temporary change is a *change* from a previous state of a certain sort.

To see why this first strategy is unsatisfactory, consider the following cases. First, consider a case where two human organisms, A and B, are identical twins and are nurtured and developed in a highly refined science laboratory from conception onwards: they are exactly similar in their genetic constitution, environmental stimuli, and so on, throughout their entire biological lives. A and B are grown up like this for many years. Each is developed to the point where she has the first-order capacity to think, but neither is developed to the point where she actually possesses the immediate capacity to think. This is because there is some neuro-physiological event, the occurrence of which is the final necessary step in the process of an organism coming to possess the immediate capacity to think, and the scientists artificially delay the occurrence of this event for A and B. Perhaps two key neurons need to communicate with one another, and the scientists insert a magnetically-activated physical barrier that functions like a gate: unless activated, the barrier remains closed, preventing the two key neurons from communicating with one another; once activated, the barrier opens up, allowing the two key neurons to communicate with one another.

So, then, A and B are both in a kind of technologically induced sleep or coma: each is highly developed enough to the point where a push of a certain button will allow her to wake up and obtain the immediate capacity to think. Now imagine that, on a certain day, both A and B are going to get their respective buttons pushed. But a double malfunction occurs. The first malfunction is that A's button works for a moment, but then stops working: perhaps the physical barrier is magnetically activated, allowing the two key neurons to begin communicating with one another, but then the physical barrier becomes de-activated, closes down, and once again prevents the two key neurons from communicating with one another. The second malfunction is that B's button gets stuck and does not work at all. The result is that A, for a moment, is allowed to develop the immediate capacity to think, but then lapses back into the state she was in before the button was pushed, whereas B is not allowed to develop the immediate capacity to think—not even for a moment. The net effect of this double malfunction is that neither A nor B have the immediate capacity to think, but A *did* actually have the immediate capacity to think—at least for a moment.

The first strategy would hold that A now has serious moral status, but B does not. But this is hard to believe. Imagine walking into the lab shortly after this double malfunction had happened, without knowing *how* it happened: that is, even though you know that one of these two had her special moment, you do not know whether it was A or B. A scientist tells you that only one of these two human organisms has serious moral status. You would be quite perplexed. After all, A and B will both develop the immediate capacity to think at the same time if they are just allowed to. It seems reasonable to think that if A has serious moral status, B also does. The mere fact that A had once possessed the immediate capacity to think should not bear the moral weight that the first strategy insists it bear.

It might be objected that this example relies upon the implausible assumption that the first strategy would be satisfied with the merest instant of possessing the immediate capacity to think. However (this objection continues), many writers hold that it is crucial whether *actual* thinking has occurred, and indeed actual thinking of a special kind: namely, self-awareness over time, accompanied by some pro-attitude, such as desire or care, that attaches to what continues over time. So (this objection concludes) our intuitions about this case would be very different indeed if A spent a period of time thinking about herself, recognizing that she is the same enduring one living, wanting that living to go on, and then sinking back into unconsciousness.

In reply, I do not think that amending this example so that A has actual self-awareness and pro-attitudes will vindicate the first strategy. However, in order to let this objection be as strong as possible, consider a very different sort of case that does not involve developing identical twins. Imagine A is a normal, healthy adult with a rich and satisfying life, endowed with the immediate capacity to think, and also endowed with self-awareness and the desire to go on living. Now imagine A undergoes a temporary change so that she is currently at t_2, and at t_2 she gets replicated in one of Derek Parfit's famous replication booths[3]: A is preserved intact and is not destroyed, but her perfect replica B is instantly produced across the laboratory. B has precisely the same sort of molecular structure that A had, and is functioning at precisely the same level as A. Furthermore, B has exactly the same capacities as A. Both A and B lack the immediate capacity to think, and both A and B will have the immediate capacity to think, along with self-awareness and the desire to go on living, at the same time if they are just allowed to.

Once again, however, the first strategy would hold that A has serious moral status, but B does not. But this is hard to believe. Imagine walking into the lab shortly after the replication had happened, without knowing *how* it happened: that is, even though you know that one of these two is a replica, you don't know whether it is A or B. A scientist tells you that only one of these two human organisms has serious moral status. You would be quite perplexed. After all, A and B will both develop the immediate capacity to think at the same time if they are just allowed to. It seems reasonable to think that if A has serious moral status, B also does. The mere fact that A had once possessed the immediate capacity to think, along with actual self-consciousness and pro-attitudes, should not bear the moral weight that the first strategy insists it bear.

In these cases involving identical twins and replicas, we can admit that there are some morally relevant differences between A and B without admitting that only A has serious moral status. For example, in the replica case, imagine that A had worked hard and put her earnings into a savings account before falling asleep and getting replicated. Now imagine that A and her replica B both come out of their sleeps, and both claim to own the money in the savings account. I believe that A has a stronger claim to the money than B, since A actually saved the money whereas B merely has pseudo-memories of saving the money. However, not all morally relevant properties are dependent like this upon the actual history of an individual. In particular, the properties that constitute serious moral status, such as the strong moral presumption against killing, do not seem to be so dependent upon an individual's actual history.

The upshot of these two cases is this. You can still have serious moral status at t_2 even if certain things were not true of you in your actual past. Thus the first strategy for replying to the temporary change argument is unsatisfactory.

3 Moral Status and the Future

A second strategy for replying to the temporary change argument is to ground your serious moral status at t_2 in the fact that certain things will be true of you in your actual future if allowed to live: for example, you actually will possess the immediate capacity to think at times like t_3, or you actually will come to have a "future like

[3] Parfit (1984, pp. 199–200).

ours" or a "future of value" because you will have conscious experiences which you will find valuable. According to this second strategy, serious moral status only exists in the case of the temporary change in virtue of the fact that the temporary change is *temporary*.

To see why the second strategy is unsatisfactory, it is important to see how it is similar to what have been called *deprivation* accounts of the wrongness of killing. Deprivation accounts of the wrongness of killing emerge from the combination of three claims:

(1) The wrongness of a particular act of killing is solely a function of the misfortune of the particular death caused by that particular act of killing.
(2) The misfortune of a particular death is solely a function of the goods that this particular death deprives the entity that dies of.
(3) The goods that a particular death deprives the entity that dies of are the goods that the entity would have had, if, contrary to fact, the entity had not died that particular death.

Such accounts of the wrongness of killing are very similar to this second strategy for replying to the problem of the temporary change. While deprivation accounts emphasize the future *goods* you would have enjoyed had you not been killed, this second strategy emphasizes the future *state* you would have been in had you not been killed. (Of course, those employing this second strategy tend to think that being in the relevant state is a *good* thing. But this is not essential to the strategy itself.)

Deprivation accounts of the wrongness of killing face what has been called a problem of over-determination.[4] For claims (2) and (3), taken together, entail something very odd whenever the following conditional holds true: if, contrary to fact, you did *not* die the particular death you *did* die, then you *would* have died some other way at the same time you actually *did* die the particular death you *did* die. Claims (2) and (3), taken together, entail that, whenever the just-mentioned conditional holds true, your particular death does not deprive you of anything and is therefore not a misfortune. When this entailment is combined with claim (1), the result is that, whenever your death is over-determined, someone can kill you without doing anything wrong.

To see why this is a problem, consider the following example. A man lives in a place where racial tensions are at the boiling point, and where race riots are known to lead to lynching that kills innocent people. He is walking down a deserted country road one night. Out of nowhere, a boy runs past him who is being chased by a mob of angry people shouting "Lynch him!" The man realizes that, no matter what he does to try and prevent it, the crowd will catch the boy and lynch him before any help arrives. In the past, the man has tried—in vain—to stop such crowds by pleas, by arguments, and by the use of force. Nothing ever works. So this time, the man

[4] See McMahan (2002, pp. 117–120).

decides to join the crowd; he himself catches up to the boy and tackles him to the ground, he himself wraps the noose around the boy's neck, and he himself pulls on the rope that hangs the boy to death from the limb of a nearby tree.

Most of us want to say that this man has done something seriously wrong. But the deprivation account of the wrongness of killing cannot say this. For even if the man had not killed the boy, the boy would have died at the same time he actually died, because someone else in the crowd would have done the same sort of thing the man did, at the same time the man did it. Therefore, the man's action did not deprive the boy of any of the goods he would have had if the action had not been performed. According to the deprivation account of the wrongness of killing, the man did not do anything wrong.

This case illustrates how the inevitability of certain sorts of outcomes is not relevant to the wrongness of certain sorts of actions. It also illustrates the inadequacy of grounding serious moral status in the fact that the individual being killed *will* be in a certain state in the future if the killing is not performed. But these illustrations are relevant to the problem of the temporary change.

Consider someone undergoing a temporary change in a situation such that they will be killed no matter what you try to do to prevent it. For example, imagine a woman undergoing anesthesia for surgery, and no matter what you do, she is going to be killed while under anesthesia: perhaps a team of utilitarian transplant surgeons is determined to remove her vital organs in order to save 5 needy patients elsewhere in the hospital. Surely, in a case like this, the woman under anesthesia still has serious moral status during her temporary change. The fact that her future will inevitably be cut short does not give us any reason to think that her serious moral status has been diminished.

The upshot is this. You can still have serious moral status at t_2 even if certain things will not be true of you in your actual future. Thus the second strategy for replying to the temporary change argument is unsatisfactory.

Of course, a hybrid strategy for replying to the temporary change argument is to combine the first two strategies; that is, to ground your serious moral status at t_2 in the fact that either certain things were true of you in your actual past (at times like t_1) or certain things will be true of you in your actual future (at times like t_3). According to this hybrid strategy, serious moral status only exists in the case of the temporary change in virtue of the fact that the temporary change is *either* temporary *or* a change from a previous state of a certain sort.

But it is hard to see how this is going to help. To see the inadequacy of this combinatorial approach, all one needs to do is to present a counterexample which combines the features of the previous counterexamples. Recall again the case where A is a normal, healthy adult with a rich and satisfying life, endowed with the immediate capacity to think, and also endowed with self-awareness and the desire to go on living. Now imagine A undergoes a temporary change so that she is currently at t_2: perhaps she undergoes anesthesia for surgery. At t_2 she gets replicated in one of Derek Parfit's famous replication booths: A is preserved intact and is not destroyed, but her perfect replica B is instantly produced across the laboratory. The replica produced, B, has precisely the same sort of molecular structure that A had, and is functioning at precisely the same level as A. Both lack the immediate capacity to think, and both will have the immediate capacity to think, along with self-awareness and the desire to go on living, at the same time if they are just allowed to. But now imagine that no matter what you do, B is going to be killed. Perhaps there is a team of utilitarian doctors who are determined to remove all of B's vital organs in order to transplant them to patients around their hospital. Whatever the reason, someone is going to kill B and nothing you can do will prevent this from happening.

The combinatorial strategy would hold that A has serious moral status, but B does not. But this is hard to believe. Imagine walking into the lab shortly after the replication had happened, without knowing *how* it happened: that is, even though you know that one of these two is a replica, you don't know whether it was A or B. A scientist tells you that only one of these two human beings has serious moral status. You would be quite perplexed. After all, both will develop the immediate capacity to think at the same time if they are just allowed to. It seems reasonable to think that if A has serious moral status, B also does. The fact that A had once possessed the immediate capacity to think, along with actual self-consciousness and pro-attitudes, conjoined with the fact that someone is going to kill B, does not give you any reason to think that B's serious moral status has been diminished.

The upshot is this. You can still have serious moral status at t_2 even if it is *both* not true that certain things were true of you in your actual past (at times like t_1), *and* not true that certain things will be true of you in your actual future (at times like t_3). Thus the combinatorial strategy under consideration is unsatisfactory as a reply to the temporary change argument.

4 Why Not Stop at the First-Order Capacity?

The temporary change argument shows that your possession of the immediate capacity to think at a given time is not the basis of your serious moral status at that time. But since you do have serious moral status at t_2, and since this serious moral status must be based upon something, what else can it be based upon? Since strategies that look to your actual past and your actual future are unsatisfactory, it seems that a satisfactory answer needs to focus on the features you actually possess at t_2.

Although you do not possess an *immediate* capacity to think at t_2, you do still possess a *first*-order capacity to think at t_2. Recall the difference between a first-order capacity and an immediate capacity. What a first-order capacity to speak English amounts to is the neurological base for speaking English, which one gets from learning the language and which one retains as long as one remains a healthy and functional organism. What an immediate or immediately exercisable capacity to speak English amounts to, on the other hand, is a first-order capacity to speak English, the exercise of which is not impeded by some transient condition. When I am awake and free from any physical impediments blocking my ability to speak, I have both a first-order and an immediate capacity to speak English. But when I am asleep, or when my mouth is full of sand, I have a first-order but not an immediate capacity to speak English.

You still have the first-order capacity to think even when you are asleep, anesthetized, or comatose. Perhaps the first-order capacity to think is both necessary and sufficient for serious moral status. If so, then it will not be necessary to accept the claim that your possession of the capacity *at some order or other* to think is the basis of your serious moral status. It will only be necessary to accept the claim that your possession of the *first*-order capacity to think is the basis of your serious moral status. The original chart above could thus be replaced with the following chart:

	t_1	t_2	t_3
Do you have serious moral status?	Yes	Yes	Yes
Do you have the immediate capacity to think?	Yes	No	Yes
Do you have the first-order capacity to think?	Yes	Yes	Yes

This reformulated chart is structurally isomorphic to the original chart. But this means that everything said above about the original chart will apply to this reformulated one as well. According to this objection, we should accept a simpler and more elegant solution to the problem of the temporary change: namely, a solution that stops at the first-order capacity to think.

What are we to say to this objection? I believe that a little reflection will show why it is a mistake to make the first-order capacity to think necessary for serious moral status. Consider again the original chart. Simply replace the question "Do you have the immediate capacity to think?" in the original chart with the question "Do you have a first-order capacity to think?" Imagine that you receive an injury which causes you to lose the first-order capacity to think for a period of time before regaining it. You retain your moral status during this period of time, even though you do not have the first-order capacity to think. Therefore, having the first-order capacity to think is not necessary for serious moral status.

Another way of putting this reply begins with the "revised" chart proposed by this objection. It is a very simple conceptual move from the *first*-order capacity to think that one possesses when asleep, anesthetized, or comatose, to the *second*-order capacity to think that is yet "higher" in terms of its order within the same hierarchy. Hence the "revised" chart proposed by this objection can be replaced with the following one:

	t_1	t_2	t_3
Do you have serious moral status?	Yes	Yes	Yes
Do you have the first-order capacity to think?	Yes	No	Yes
Do you have the second-order capacity to think?	Yes	Yes	Yes

This chart is structurally isomorphic to both the original one and the revised one proposed by this objection. But this means that everything said above about the original chart and the revised chart will apply to this chart as well. The original chart and the revised chart showed how the possession of the immediate capacity to think was not necessary for you to have serious moral status. This chart shows how the possession of the first-order capacity to think is not necessary for you to have serious moral status.

Notice, however, that what is true about the first-order capacity to think is also true about any given order of the capacity to think. For any numerically-characterized order of the capacity to think (e.g., the 100th-order), we can always imagine a temporary change scenario in which that particular order of the capacity to think is lost, a higher-order capacity to think (e.g., the 101st-order) remains, and your serious moral status remains. Once you start down the path of higher-order capacities, there is no way—or at least no principled way—of turning back. The structure of a temporary change can be utilized on any two adjacent capacities in a hierarchy of capacities:

	t_1	t_2	t_3
Do you have serious moral status?	Yes	Yes	Yes
Do you have the *n*th-*order* capacity to think?	Yes	No	Yes
Do you have the (*n+1*)th-*order* capacity to think?	Yes	Yes	Yes

Thus the structure of the temporary change argument can be iterated as long as there is a higher-order capacity to think. It would seem, then, that the temporary change argument is able to show that the fact that you possess *some* higher-order capacity to think is sufficient for you to have serious moral status. In other words, your capacity *at some level or other* to think is what generates your serious moral status.

If the moves of this chapter up until this point are sound, then it seems to follow that normal human infants have serious moral status. For the sake of simplicity, let us again focus on the capacity to think. Notice what follows from the fact that your possession of the capacity *at some order or other* to think is sufficient for you to have serious moral status. As long as you already have the capacity (at some order or other), you already have serious moral status *even before* your very first moment of possessing the immediate capacity to think.

So, one of the philosophical moves above is from the first-order capacity to think to some higher-order capacity to think, and another philosophical move focuses on the times of your life before you possess the immediate capacity to think for the first time. When these two moves are combined, they seem to provide a promising way of reasoning with philosophers who do not begin with the intuition that normal human infants have serious moral status. What the two moves show, when combined, is that as long as you have some higher-order capacity to think during the time of your infancy, then you have serious moral status during the time of your infancy.

At this point, it is worth pausing to mention a few objections to my attempt to prove the serious moral status of normal human infants. First, why should someone believe that a human infant actually has *any* higher-order capacity to think? Those asking this question could *agree* with the claim that your possession of a higher-order capacity to think is sufficient for you to have serious moral status, even before your very first moment of possessing the immediate capacity to think. But they could *disagree* with the claim that you possessed this higher-order capacity when you were an infant, on the grounds that you *never were* an infant.

Since I already answered this objection in Chapter 2, I shall merely repeat the conclusion of that answer here. "Metaphysical" temporary change cases, like the cases of Albert and Benjamin, illustrate that you could become the capacity-equivalent of a human infant. You could become infantilized. Since you could become an infantilized human adult like Albert, there is no reason to think you could not have been a normal human infant like Benjamin. Indeed, once we realize that an adequate theory of personal identity must make room for you to become infantilized like Albert, there is every reason to believe that you actually were a human infant at one time, possessing merely the higher-order capacities, back then, to do the things you are in fact doing, right now.

I would like to focus on a related but distinct objection: why should someone believe that a human infant possesses the *sort* of higher-order capacity to think that generates serious moral status? Those making this objection could *agree* with the claim that your possession of a higher-order capacity to think is sufficient for you to have serious moral status, even before your very first moment of possessing the immediate capacity to think. But they could *disagree* with the claim that you possessed this higher-order capacity when you were an infant, on the grounds that the *order* of this capacity when you were an infant was *too high* to generate serious moral status. Perhaps there is an order in the hierarchy of capacities related to thinking—perhaps the 3rd or 4th order, perhaps the 30th or 40th—where serious moral status is no longer generated. After all, even if it is true that you once were an infant, and even if it is true that your serious moral status *when sleeping* is generated by the higher-order capacity to think you possess when sleeping, it still needs to be established that you had moral status *when an infant* because of the higher-order capacity to think you possessed *when an infant*. For the capacity to think you possessed when an infant is of a different and indeed much higher order than the capacity to think you possess when sleeping. Those making this objection might believe that *some* higher-order capacities to think generate serious moral status, while *other* higher-order capacities to think do not.

This objection will now be answered by a further argument, whose basic thrust can be summarized in this paragraph and whose premises can be set out and defended in the next few paragraphs. As we imaginatively alter the temporary changes you could undergo as an adult, the order of the capacities we must appeal to, in order to generate your serious moral status, gets higher and higher. Since previous possession of the immediate capacity to think is irrelevant, it follows that as we heighten the order of capacities sufficient to generate serious moral status for an individual *who has possessed but then lost* the immediate capacity to think, we thereby heighten the order of capacities sufficient to generate serious moral status for an individual *who has not already possessed* the immediate capacity to think. Eventually, we reach a point where the order of the capacity to think possessed by the individual in the middle of a temporary change is the same as the order of the capacity to think possessed by an infant: for example, just as an infant only has a (say) 100th-order capacity to think, so too we can envisage an adult in the middle of a temporary change who only has a 100th-order capacity to think. But since possession of this capacity is enough to generate serious moral status for the adult, possession of this capacity is also enough to generate serious moral status for the infant.

More formally, this "comparison" argument can be set out as follows:

(1) Some adults are the same, in terms of the order of their capacity to think, as some infants.
(2) For the adults in premise (1), having their particular order of the capacity to think is sufficient to generate serious moral status.
(3) For any individuals x and y, for any activity A, and for any order N of the capacity to A, if x's having the N-th order capacity to A is sufficient to generate serious moral status for x, then y's having the N-th order capacity to A is sufficient to generate serious moral status for y.

Therefore,

(4) For the infants in premise (1), having their particular order of the capacity to think is sufficient to generate serious moral status.

I already defended premise (1) in Chapter 2, with the comparison between Albert and Benjamin. And it is not hard to see that premise (2) can be defended as well. Surely, Albert still has serious moral status during the period of his temporary change. Otherwise, certain sorts of temporary injuries or setbacks can cause your serious moral status to disappear. Furthermore, it can be shown (via the arguments above) that Albert's serious moral status is generated by Albert's possession of his higher-order capacity to think, and not, for example, by the fact that Albert once possessed the immediate capacity to think. The case of Albert and Benjamin therefore supports premise (2) in the above argument as well.

The fact that the temporary change argument can be iterated is what allows me to argue that human infants have serious moral status.

Likewise, the fact that the temporary change argument can be iterated is what allows me to argue that human fetuses, human embryos, and all manner of "marginal cases" have serious moral status. As we imaginatively alter the temporary changes, the order and type (active or passive) of the capacity to think possessed by the individual in the middle of a temporary change become the same as the order and type of the capacity to think possessed by a human fetus, or a human embryo, or a "marginal case." But since possession of this order and type of capacity is enough to generate serious moral status for the former, possession of this order and type of capacity is also enough to generate serious moral status for the latter. All the comparison cases from Chapter 2—Albert and Benjamin, Caleb and Drake, and so on—reappear, in the context of serious moral status, to show that if the first member of the comparison has serious moral status, then so does the second.

Imaginary temporary change cases therefore explain why it is not enough to base serious moral status on neurophysiological features of a human organism like the fact that it has a functioning cerebral cortex. To see why, simply replace the question "Do you have the immediate capacity to think?" in the original chart with the question "Do you have a functioning cerebral cortex?" Imagine that you receive an injury to your cerebral cortex, which causes that organ to cease functioning for a period of time before it resumes functioning. You retain your moral status during this period of time, even though you do not have a functioning cerebral cortex.

One objection should be briefly mentioned. One group of philosophers, though willing to admit that you could have been an infant or fetus (or embryo or zygote), still deny that you had serious moral status at that time. When confronted with the case of an adult human organism that has serious brain damage, some of these philosophers, though willing to admit that you could survive such damage, still deny that you would have serious moral status at that time. For example, David Boonin writes, at the outset of his book *A Defense of Abortion,* that

> In the top drawer of my desk, I keep another picture of [my son] Eli. This picture was taken on September 7, 1993, 24 weeks before he was born. The sonogram image is murky, but it reveals clearly enough a small head tilted back slightly, and an arm raised up and bent, with the hand pointing back toward the face and the thumb extended out toward the mouth. There is no doubt in my mind that this picture, too, shows the same little boy at a very early stage in his physical development. And there is no question that the position I defend in this book entails that it would have been morally permissible to end his life at this point.[5]

Although an examination of the details of Boonin's own position is beyond the scope of this book, a move he makes when defending that position is directly relevant to the points being made in this chapter. Boonin considers a counterexample to his view of "...an imaginary case in which a temporarily comatose adult has had the entire contents of his brain destroyed so that there is no more information contained in his brain than is contained in that of the preconscious fetus." Boonin argues that, in a case such as this,

> It seems right that my position does not imply that such an individual has the same right to life as you or I. But ... a critic of abortion cannot appeal to such a case as a means of rejecting my position because we cannot assume ahead of time that killing such individuals is seriously immoral.[6]

The last claim that Boonin makes in this passage—that we cannot assume ahead of time that killing certain sorts of temporarily comatose human individuals is seriously immoral—is a substantive, and surprising, claim.

I believe that the last claim Boonin is making here is simply mistaken. At this particular point in his argument, Boonin has misidentified who needs to bear the burden of proof. We are well within our rights to assume, ahead of time, that killing such individuals is seriously immoral. Any one of us could become such a temporarily comatose adult. If a cure for our condition were available—remember the case

---

[5]Boonin (2003, p. xiv).

[6]Boonin (2003, p. 78).

of Ronald Reagan at the beginning of this book—it would not be a matter of moral indifference whether we were killed or cured.

So, the assumption that killing such individuals is seriously immoral is an assumption we are entitled to use in framing our moral views. Someone who denies that killing such individuals is seriously immoral, on the other hand, needs to have a convincing argument for this denial. And this is so, regardless of what our current views are about the morality of abortion. If a given moral view cannot accommodate this assumption, this is a mark against the view and not a mark against the assumption.

So, then, the result of this temporary change argument is that if you have the capacity, at some order or other, to think, you have serious moral status. But of course, there is no reason to think this is unique to *you*: the temporary change argument could apply to any entity. Also, it was noted above that the capacity to think was chosen merely for the sake of simplicity: the temporary change argument could focus on the entire set of immediate capacities possessed by any normal adult human person. It would seem, then, that the temporary change argument is able to show that the fact that an entity possesses a set of typical human capacities, at some level or other, is sufficient for that entity to have serious moral status. And this, of course, is the very claim this chapter is seeking to prove.

5 Actual, Continuing Subjects of Experience

The view that has been defended in this chapter can be expressed as follows:

(C) If something has a set of typical human capacities, it has serious moral status.

If the defense of (C) is to be successful, then one of the alternative views that needs to be refuted is the following one:

(P) The only things that have serious moral status are actual, continuing subjects of experiences.

I would like to briefly explain (P) and highlight how (P) differs from (C). Then, I would like to explain why one of the reasons sometimes given for accepting (P)—namely, that it adequately handles thought-experiments that involve what may be called "complete reprogramming" of the upper brain—is actually a reason for favoring (C) over (P).

Chances are, everyone reading this sentence is an actual, continuing subject of experiences. But to understand what it means to be an actual, continuing subject of experiences, it is important to emphasize that a *continuing* subject of experiences is different than a *momentary* subject of experience. Suppose that there is an organism A that can experience pleasure, but that has no thoughts at all, including no memories of past experiences, no intentions for obtaining future experiences, and so on. One moment, A is enjoying the pleasure of eating. The next moment, A is enjoying the pleasure of standing in the shade. The next moment, A is enjoying the pleasure of resting. One might be forgiven for thinking that A itself is a continuing subject of experiences. But this is one of the things that are denied by those who employ the concept of actual, continuing subjects of experience. According to them, there is no *continuing* subject of experiences and other mental states associated with A, but only a series of psychologically isolated *momentary* subjects of experience.

Suppose next that there is an organism B that has thoughts, apparent memories, and intentions regarding the future, but that has these mental states in a way that appears completely random from one moment to the next. One moment, B is beginning the thought "I think, therefore, I am." But before B can finish this thought, B's mental state suddenly switches to the momentary belief that it enjoyed visiting Beijing. The next moment, B's mental state switches to the intention to read Paul's letter to the church at Rome. And so on. As with organism A, there is no continuing subject of experiences and other mental states associated with B, but only a series of psychologically isolated momentary subjects of experience.

When, then, does an organism have a continuing subject of experience associated with it? The short answer is: when the organism has certain sorts of mental states that are causally connected to one another in the right sorts of ways. And what this usually amounts to, for those who take this approach, is that the organism needs to have certain sorts of *brain* states that are causally connected to one another in the right sorts of ways. Of course, it is possible, in principle, to believe that some angels or immaterial minds could be, or could at least be associated with, actual, continuing subjects of experiences. But those who speak of actual, continuing subjects of experiences usually hold the view that experiences themselves, momentary subjects of experiences, and actual, continuing subjects of experiences all depend upon certain parts of the brain. The result is that, if an actual, continuing subject of experiences is associated with a given human organism, then once the relevant brain regions of that human organism are damaged or destroyed, the actual, continuing subject of experiences ceases to exist.

The view that actual, continuing subjects of experiences have serious moral status is able to handle some of the temporary change scenarios envisioned above at least as well as the view that possessing a set of typical human capacities suffices to give an entity serious moral status. After all, when you are asleep, anesthetized, or comatose, the relevant brain states that give rise to an actual, continuing subject of experiences presumably remain intact. However, some of the other scenarios envisaged above can only be seen as a direct challenge to the idea that *only* actual, continuing subjects of experiences have serious moral status. This is because, in order to be an actual, continuing subject of experiences, it is necessary to have actually had some of the right sorts of experiences already, in the past, and to have those past experiences causally connected to other sorts of experiences in the right sorts of ways. So, for example, in the replica cases, since the original, but not the replica, has an actual, continuing subject of experiences associated with it, it follows that the original, but not the replica, has serious moral status.

One of the more important things to emphasize in this context is this: the mere fact that a human organism has certain sorts of brain states does not by itself entail that the organism has an actual, continuing subject of experiences associated with it. Imagine that the replica cases envisioned above had not involved a temporary incapacitation of the original A: A gets replicated, let us say, while she is wide awake and enjoying her life very much. The replica produced, B, does not have an actual, continuing subject of experiences associated with it until B actually begins having the relevant sorts of experiences, and these experiences are causally related in the rights sorts of ways. So even after B has her first experience (such as the thought "am I the original or am I the replica?"), this only means that a *momentary* subject of experiences is associated with B; a *continuing* subject of experiences does not exist until it can be built up out of at least two momentary subjects of experiences, causally related in the right sorts of ways.

One of the reasons sometimes given for accepting (P) is that it handles the following sort of case very well, and indeed much better than its rivals:

> Suppose . . . that there are technological developments that allow the brain of an adult human to be completely reprogrammed, so that the organism winds up with memories (or rather, apparent memories), beliefs, attitudes, and personality traits completely different from those associated with it before it was subjected to reprogramming. (The pope is reprogrammed, say, on the model of David Hume.) In such a case, however beneficial the change might be, most people would surely want to say that *someone* had been destroyed, that an adult human being's 'right to life' had been violated, even though no biological organism had been killed.[7]

Many people have the intuition that such "complete reprogramming" of a normal adult human organism's brain is morally on a par with killing that human organism. (P) apparently explains these intuitions, but these intuitions receive no explanation at all given (C), since reprogramming neither kills any human organism, nor does it change the typical human capacities that the human organism in question has.

I believe that the most direct way of replying to this argument is to show that careful consideration of cases that involve complete reprogramming actually provides one with several reasons for favoring (C) over (P). I will now consider several related cases, explaining why (C) handles these cases better than (P). For convenience, I will speak in terms of "serious moral status" instead of "a right to life", "a strong moral presumption against killing," and "a right to continued existence".

Case 1: *Malicious Reprogram, No Experience Yet*: Imagine a malicious reprogrammer reprograms the brain of George W. Bush at some time t so that it (the brain) gives rise at t* to the exact same types of beliefs and desires as John Kerry's brain. But at t* the reprogrammed brain has not given rise to any actual experiences. (C) Maintains that the human organism with this reprogrammed brain still has serious moral status at t*. But (P) denies this. (P) is mistaken; the human organism does still have serious moral status at t*.

---

[7] Tooley (1983, pp. 102–103).

Case 2: *Malicious Reprogram, No Experience Yet, Benevolent Reprogram.* Case 2 is the same as Case 1, only now there is a benevolent reprogrammer who can reverse the effect of the malicious reprogrammer at t* so that the brain gives rise at t** to the exact same types of beliefs and desires as George W. Bush originally had at t. (C) maintains (1) that the human organism with the reprogrammed brain still has serious moral status at t*, (2) that there is excellent reason to let the benevolent reprogrammer reprogram the brain at t*, (3) that the human organism with the re-reprogrammed brain still has serious moral status at t**. But (P) denies all three of these claims, maintaining instead (1) that the human organism with the reprogrammed brain does not have serious moral status at t*, (2) that it is a matter of indifference whether the benevolent reprogrammer reprograms the brain at t*, and (3) that the human organism with the re-reprogrammed brain still does not have serious moral status at t**. But all three of these holdings of (P) are mistaken.

Case 3: *Malicious Reprogram, Experience, Benevolent Reprogram.* Case 3 is the same as Case 2, only now, let the human organism with the reprogrammed brain have some appropriately related experiences between t* and t+, which is the time at which the benevolent reprogrammer arrives on the scene. (C) maintains that there is excellent reason to let the benevolent reprogrammer reprogram the brain at t+. But (P) claims that it would be morally wrong for the benevolent reprogrammer to reprogram the brain at t+; this, (P) claims, would be morally akin to murder. But this is a mistake; reprogramming at t+ is not morally akin to murder: in fact, it is the morally proper thing to do, since it is simply reversing the effect of the malicious reprogrammer, and giving back to George W. Bush the distinctive personality which he had at t.

Case 4: *Double Malicious Reprogram, No Experience Yet.* Case 4 is the same as Case 1, only this time the malicious reprogrammer reprograms the brains of both George W. Bush and John Kerry at t so that they (the brains) give rise at t* to the exact same types of beliefs and desires that the other person's brain gave rise to at t. The original Bush brain gives rise at t* to the beliefs and desires the original Kerry brain had at t; the original Kerry brain gives rise at t* to the beliefs and desires the original Bush brain had at t. But at t* neither reprogrammed brain has given rise to any actual experiences. (C) maintains that the human organisms with these reprogrammed brains still have serious moral status at t*. But (P) denies this. (P) is mistaken; the human organisms do still have serious moral status at t*.

We could easily construct additional cases that parallel Case 2 and Case 3 in the same sorts of way Case 4 paralleled Case 1. A case paralleling Case 2 may be called *Double Malicious Reprogram, No Experience Yet, Benevolent Reprogram.* The case paralleling Case 3 may be called *Double Malicious Reprogram, Experience, Benevolent Reprogram.* But I think that it would be more instructive to leave behind for the moment malicious reprogrammers bent on altering the brains of George W. Bush and John Kerry, and focus instead on a case where a human organism might really want to have its brain reprogrammed.

So, then, Case 5 may be called *Repentant Man, Benevolent Reprogrammer*: Imagine a mature adult human organism comes to the conviction that most of his central beliefs about himself and his place in the world are false, and that most of his deepest desires are wicked. He wants to change his personality so that he becomes like Gandhi, but he realizes that the project of such self-transformation will take him more time than his biological organism has left to live. However, this man knows a reprogrammer who can reprogram his brain so that it gives rise to the sorts of beliefs and desires that Gandhi had. (C) does not automatically praise or condemn such reprogramming. On the one hand, if the repentant man asked the reprogrammer to do it, it might be morally permissible for the reprogrammer to reprogram the repentant man in the way he asks, since the repentant man would persist through the change and would himself be the beneficiary of the change. On the other hand, it is possible to think that such reprogramming would be wrong, since it may be a form of "cheating" similar to the taking of illegal substances to enhance one's performance in a sport: perhaps a character like Ghandi's is only valuable if the agent who has it was directly responsible for bringing it about.[8] In any event, regardless of whether (C) would fit better with praising or condemning such reprogramming, the relevant point for our purposes is that (P) seems to take an entirely different view of this case. In the first place, (P) would insist that the repentant man is really requesting something akin to annihilation, or assisted suicide. According to (P), whether the repentant man realizes it or not, "he" is going to be gone forever once the reprogramming happens. Perhaps (P) will not condemn the repentant man's decision, since (P) takes no stand on whether an actual, continuing subject of experiences has the right to wave its right to continued existence. But (P) will insist that the repentant man does not continue to exist past the point of reprogramming. In the second place, however, according to (P), the human organism that does exist right after the reprogramming does not have serious moral status until it actually begins having some experiences. But both of these implications of (P) seem strained.

Finally, it is worth noting that (C) is better able than (P) to handle other sorts of cases, which involve, not the reprogramming of a human organism's brain, but the temporary destruction of a human organism's upper brain. Case 6 may be called *Voluntary Operation*. Imagine a mature adult human organism finds out from his doctors that he has a degenerative disease in his upper brain that will eventually result in his biological death, and that the only way to prevent this from happening, since the disease has spread throughout his upper brain, is to agree to an operation that involves both destroying, and then reconstituting, the upper brain itself. One way this might work is as follows: a structural snapshot of his upper brain is "scanned out" into a computer file, but the process of "scanning out" necessarily involves the complete de-programming of the upper brain tissues themselves; the computer file is altered so that the parts of the file corresponding to the degenerative disease are removed; the altered structural snapshot is "scanned in" to his de-programmed upper brain so that the brain has the exact same structure that it had at first, minus the disease. Another way this might work is this: a surgical vacuum cleaner is used to painlessly suck out and transport all the upper brain tissues from the man's skull to a "cleaning tank" where the diseased tissue is separated from the

---

[8] Thanks to Chris Tollefsen for pressing me on this point.

healthy tissue via a process of high-speed spinning, and the healthy tissue is then put back into the man's skull with the same structure as it had before, minus the damage done by the disease. In any event, (C) has no objection to these procedures: the man being cured is there at the beginning of the operation, there in the middle of the operation, and there at the end—and he retains his serious moral status throughout the operation. But (P) denies that the man is there in the middle of the operation, denies that the man is there at the end of the operation, denies that the human organism in the middle of the operation has serious moral status, and denies that the human organism at the end of the operation has serious moral status. (P) seems to me to be mistaken in all four of these denials.

The same sort of case can be constructed without the voluntariness of the patient, and without the disease to eliminate. Consider Case 7: *Involuntary Experiment*: Imagine the following alien abduction story. It is a "kinder, gentler" alien abduction story, partly because it all happens while the abductee is asleep, so there is no experiential suffering on his part. A kind and gentle alien removes, and completely destroys, the upper brain of a normal adult human organism, but leaves the rest of the organism alive. It puts all the parts of the upper brain into an alien experiment and totally disaggregates them, right down to the simplest atoms. But then the kind and gentle alien re-assembles those parts, creating a new upper brain, indistinguishable from the original one in every way: the new upper brain not only has the same structure as the original upper brain, but has, in addition, the exact same parts in all their exact old places: atom #4321 is next to atom #4322, just as before, and so on. Finally, the alien implants the "new" upper brain where the "old" upper brain was, and flies away. All this happens without waking up the human organism in question. According to (C), the man being experimented on is there at the beginning of the experiment, there in the middle of the experiment, and there at the end—and he retains his serious moral status throughout the operation. But (P) denies that the man is there in the middle of the experiment, denies that the man is there at the end of the experiment, denies that the human organism in the middle of the experiment has serious moral status, and denies that the human organism at the end of the experiment has serious moral status. (P) seems to me to be mistaken in all four of these denials.

The basic points that emerge from these cases are as follows. Sometimes reprogramming a human organism's brain is positively the right thing to do. Furthermore, the fact that there is not an "actual, continuing subject of experiences" associated with an organism does not mean that the organism is lacking in serious moral status. Consequently, these cases support (C) over (P).

It is time to sum up the discussion of the first five sections of this chapter. I have argued that there are cases of human organisms that are in such a state that the best available basis for their serious moral status is their set of typical human capacities. Since most of us—philosophers and non-philosophers alike—do believe that these human organisms have serious moral status, and since most of us do base this serious moral status on something, it must be the set of typical human capacities that we base it on. The human organisms in question were those who undergo various "temporary changes" like the temporary changes described in Chapter 2.

I then considered four other available features of yours, besides the feature of having a capacity (at some order or other) to think, that might be thought satisfactory for generating serious moral status: (A) the property of having a certain sort of past; (B) the property of having a certain sort of future; (C) the property of having the first-order capacity to think; (D) the property of being an "actual, continuing subject of experiences." I argued that none of these properties, whether considered jointly or together, are satisfactory as properties for basing your possession of serious moral status on.

The surprising result of this "moral" temporary change argument is that if something has a set of typical human capacities, it has serious moral status. Up until this point in the chapter, I have relied, not on moral theory, but on moral intuitions about temporary change cases. However, I shall now conclude this chapter by arguing that the same surprising result can be arrived at by adopting the moral framework of either John Rawls or Martha Nussbaum.

6 Capacities and the Original Position

Two different discussions in *A Theory of Justice*—one on the basis of equality, the other on paternalism—lead naturally to the idea that if something has a set of typical human capacities, it has serious moral status.[9] In broad outline, Rawls' discussions on these two topics lead naturally to this idea in the following ways. On the topic of paternalism, Rawls discusses how the parties in the original position would seek to protect themselves in case they end up as incapacitated or undeveloped human beings when the veil of ignorance is lifted. Hence the parties to the original position have every reason to agree with the claim that if something has a set of typical human capacities, at some order or other, then it has serious moral status. On the topic of the basis of equality, Rawls discusses how the parties in the original position would guarantee basic rights for all those with the *capacity* to take part in this original position. In order to guarantee the basic rights of infants and young children, he goes on to interpret this capacity as a "potentiality that is ordinarily realized in due course." In other words, the part of serious moral status that falls under the basic rights of individuals is based on the possession of certain higher-order capacities typical of normal adult human organisms.

Let us examine these Rawlsian arguments in more detail, starting first with his discussion of paternalism in *A Theory of Justice*. Rawls discusses how the parties in the original position would seek to protect themselves in case they end up as incapacitated or undeveloped human beings when the veil of ignorance is lifted:

> In the original position the parties assume that in society they are rational and able to manage their own affairs... But once the ideal conception is chosen, they will want to

---

[9] What follows is adapted from DiSilvestro (June 2005).

> insure themselves against the possibility that their powers are undeveloped and they cannot rationally advance their interests, as in the case of children; or that through some misfortune or accident they are unable to make decisions for their good, as in the case of those seriously injured or mentally disturbed... For these cases the parties adopt principles stipulating when others are authorized to act in their behalf and to override their present wishes if necessary; and this they do recognizing that sometimes their capacity to act rationally for their good may fail, or be lacking altogether.[10]

The following argument seems implicit in this passage:

(1) The parties in the original position are concerned to protect themselves during all the times of their lives when their capacity to act rationally for their good is *undeveloped* [or failing, or lacking altogether].

(2) For the parties in the original position, the times of their lives when they are in a *childhood* state [or an injured state, or a foolish state] are among the times of their lives when their capacity to act rationally for their good is undeveloped [or failing, or lacking altogether].

Therefore,

(3) The parties in the original position are concerned to protect themselves during the times of their lives when they are in a *childhood* state [or an injured state, or a foolish state].

This argument's relevance to higher-order capacities can be made clear by considering what sorts of ideas one is committed to in assuming the truth of premise (2). First, it seems that (2) involves some sort of understanding of *capacities*: a party to the original position can imagine her *capacity* to act rationally for her good being undeveloped, etc. How might this be filled in? For starters, if x (e.g., a child) has the capacity to ϕ (e.g., to act rationally for her own good), then x has this capacity even when x is not ϕ-ing. Thus the mere fact that x is not *currently* ϕ-ing does not threaten the claim that x has the capacity to ϕ. This helps explain why someone can have the capacity to act rationally for her good even though that capacity (as Rawls puts it) *may fail*.

Second, it seems that premise (2) involves some sort of understanding of *personal identity*. The parties in the original position are not concerned to protect *other* people but *themselves* in case *they* end up in an unfortunate condition; this presupposes that the rational persons in the original position are the *same persons as* the irrational persons they can imagine becoming. How might this be filled in? For starters, if x has the capacity to ϕ, then x must be the same thing when she is ϕ-ing as when she is not ϕ-ing. Without this sameness, it would not really be x that has the capacity to ϕ but some other thing y. Indeed, without this sameness, this other thing y would not really have the capacity to ϕ; what y would have is some sort of claim to be the "ancestor" of the x that actually does the ϕ-ing.

---

[10] Rawls (1971, pp. 248–249).

Third, it seems that premise (2) involves some sort of understanding of what it is for a capacity to belong to someone even when that capacity is *undeveloped* or *lacking altogether*. Both of these notions can be clarified by the following idea: x's capacity to ϕ, although in a real sense still *belonging to* x, need not be *immediately exercisable by* x. In Rawls' example of children, a child's capacity to act rationally for her good, while properly belonging to the child, might nevertheless be the sort of thing we must wait on before we can see "in action." Part of the idea of an undeveloped capacity is that *once certain conditions are met*, x will have the *immediately exercisable* capacity to ϕ. In the case of children, such conditions include a certain amount of biological growth and maturation; when these conditions of growth and maturation are met, the x that started off as a child will eventually have the immediate capacity to act rationally for her good.

Another way of putting this idea is that having an undeveloped capacity to ϕ is like having a *capacity* for having the *first*-order capacity to ϕ. The capacity to *have* this first-order capacity can be understood as a *second-order* capacity. The most generic way of stating this idea is that capacities come in *hierarchies* of lower and higher orders.

To see why Rawls is assuming this hierarchical picture, recall how the *failure to realize* a capacity is not the same thing as the *non-existence of* a capacity. Just as the mere fact that x is not currently ϕ-ing does not threaten the claim that x has the capacity to ϕ, so too the mere fact that x *lacks* the first-order capacity to ϕ does not threaten the claim that x has the second-order capacity to have this first-order capacity. This helps us make sense out of Rawls' claim that the parties in the original position are concerned about how they will be treated when their capacity to act rationally for their good is (as Rawls puts it) *lacking altogether*. Presumably this means that they are lacking an immediate capacity to act rationally for their good, or perhaps that they are lacking even a first-order capacity to act rationally for their good; but that *were certain conditions to be met*, they would have this immediate capacity, or they would have this first-order capacity. For example, if someone with a mental disease were unable to think straight due to a certain imbalance of chemicals in the brain, although in one sense she would be *lacking altogether* the capacity to act rationally for her good—in other words, she lacks the immediate capacity to act rationally for her good, and perhaps even lacks the first-order capacity to act rationally for her good—nevertheless, *were this imbalance to be corrected for*, she would have the immediate capacity (or at least the first-order capacity) to act rationally for her good. If this is correct, then perhaps Rawls' claim that "sometimes their capacity to act rationally for their good may . . . be lacking altogether" could be read instead as "sometimes their capacity to act rationally for their good may be *seriously undeveloped*."

Rawls himself uses the notion of a hierarchy of capacities in describing one of the two features of persons in *Political Liberalism:*

> . . .since persons can be full participants in a fair system of social cooperation, we ascribe to them the two moral powers connected with the elements in the idea of social cooperation noted above: namely, a *capacity* for a sense of justice and a capacity for a conception of the

> good. A sense of justice is the *capacity* to understand, to apply, and to act from the public conception of justice which characterizes the fair terms of social cooperation...[11]

The italics are mine to highlight the following: since a sense of justice *just is* a certain sort of capacity (namely, a capacity to understand, to apply, and to act from the public conception of justice...), then the capacity *for* a sense of justice *just is* a capacity for a capacity. What makes you a member of a Rawlsian moral community is a capacity for a capacity.

This elaboration about hierarchies of capacities leads naturally to a fourth idea that must be assumed for premise (2) to make sense, and this idea again involves personal identity. It seems that *x* must be the same thing when she has *merely* the higher-order capacities as she is when she has *both* the higher and the lower-order capacities. Without this sameness, some other thing *y* would have *merely* the higher-order capacities, and *y* would be some sort of "ancestor" to the *x* that has *both* the higher and the lower-order capacities. With this sameness, however, it makes perfect sense to claim that a rational adult could thank you for treating her paternalistically when she was an irrational child.

Why might Rawls be justified in assuming the child is the same person as the adult she grows and develops into? Since personal identity is more of a metaphysical question than a purely political one, Rawls deliberately tries to avoid making controversial claims about it.[12] Still, any account must make certain assumptions about personal identity, and Rawls' account is no exception. I propose, negatively, that the *ground* of personal identity between children and adults Rawls is assuming could not be a matter of the *bodies* of children and adults being composed of the same materials, for this is never the case due to the well-known molecular turnover in the human body which occurs every several years. The cells that compose the child are not the same cells that compose the adult, but this provides us no challenge to our view that the child and the adult are literally the same being existing at different times. Nor is *memory* an adequate ground of the identity between child and adult, since few adults remember much of their childhood, despite the fact that it was still *their* childhood they have forgotten. Finally, this ground is not to be found in something like a *similarity of fundamental beliefs, projects, and attitudes about life*. Although a few people may have similarities of this sort between childhood and adulthood, most do not. So the ground of personal identity Rawls is assuming must be something else. I propose, positively, that Rawls is assuming that this personal identity is grounded in something like *spatiotemporal or causal continuity* between the child and the adult. The best candidate of this sort, and the one Rawls is probably assuming, is that the child and the adult are same *organism*.

The relevant point, however, is that *lacking* the immediate capacity to act rationally for one's good cannot disqualify someone from being an object of concern for the parties in the original position. The parties want to look out for the well-being of individuals who lack this immediate capacity, because in so doing, the parties

---

[11] Rawls (1993, p. 19).

[12] Rawls (1993, p. 32), footnote 34.

are looking out for their *own* well-being. Put differently, in terms of serious moral status, the parties in the original position would agree that they have serious moral status, and they would seek to formulate principles of justice that recognized the presence of serious moral status in individuals who lack the immediate (or even-first-order) capacity to act rationally for their good. Hence the parties to the original position have every reason to agree with the claim that if something has a set of typical human capacities, at some order or other, then it has serious moral status. This, then, is the relevance of Rawls' discussion of paternalism to the claim this chapter has been defending.

On the topic of the basis of equality, Rawls discusses how the parties in the original position would guarantee basic rights for all those with the *capacity* to take part in this original position. In order to guarantee the basic rights of infants and young children, he goes on to interpret this capacity as a "potentiality that is ordinarily realized in due course." Let us examine this discussion of the basis of equality more closely.

Broadly, Rawls thinks the basis of equality consists of those features of humans beings "in virtue of which they are to be treated in accordance with the principles of justice." By clarifying these features, Rawls wants to address the grounds on which "we distinguish between mankind and other living things and regard the constraints of justice as holding only in our relations to human persons."[13]

Rawls distinguishes three levels of applicability for the concept of equality. The first and least controversial level is "justice as regularity", exemplified by the impartial application of rules (e.g., "treat similar cases similarly"). The second and more difficult level is the basic structure of society, where equality gets specified by the two principles of justice, which in turn demand "that equal basic rights be assigned to all persons." The third level is where we specify what "persons" *are*; or, the way Rawls puts it, "what sorts of beings are owed the guarantees of justice." Until this third level is explored, both the inclusion of humans among those "sorts of beings" and the exclusion of animals from those "sorts of beings" are left unexplained.[14] His approach to this third level is that the only "persons" who are entitled to equal justice are "*moral* persons" distinguished by two features:

> first they are capable of having (and are assumed to have) a conception of their good (as expressed by a rational plan of life); and second they are capable of having (and are assumed to acquire) a sense of justice, a normally effective desire to apply and to act upon the principles of justice, at least to a certain minimum degree.[15]

Moral persons are thus characterized by their *capacities* (as is evident from the twice-repeated locution "they are *capable* of having"), and Rawls' very next move is to discuss the link between these capacities and the deliberations of the parties in the original position. Since the parties to the original position adopt the principles they do "to regulate *their* common institutions and *their* conduct toward one

---

[13]Rawls (1971, p. 504).

[14]Rawls (1971, p. 505).

[15]Rawls (1971, p. 505).

another," and since the reasoning behind the selection of these principles factors in "the description of *their* [the parties'] nature," it follows that "equal justice is owed to those who have the *capacity* to take part in and to act in accordance with the public understanding of the initial situation."[16]

An important question now becomes: is this *capacity*, which characterizes moral persons, best understood as an immediate capacity, or a first-order capacity? Or is it better understood as a *higher-order* capacity? Rawls' very next comment clearly establishes the latter understanding:

> ...moral personality is here defined as a potentiality that is ordinarily realized in due course. It is this potentiality which brings the claims of justice into play. I shall return to this point below.[17]

This potentiality is precisely the sort of higher-order capacity we looked at in the context of Rawls' discussion of paternalism. To confirm this, we need only look at the passage Rawls refers to in his promissory note to "return to this point below." When he returns to this point, he cites the need to guarantee the rights of children and infants as one of the motivations for construing the basis of equality in just the way he construes it:

> ...the minimal requirements defining moral personality refer to a capacity and not to the realization of it. A being that has this capacity, whether or not it is yet developed, is to receive the full protection of the principles of justice. Since infants and children are thought to have basic rights ... this interpretation of the requisite conditions seems necessary to match our considered judgments.[18]

What is clearly doing the work here is a notion of *higher-order* capacities possessed by the infants and children.

Before examining the rest of Rawls' discussion of the basis of equality, then, we can suggest a preliminary analysis for how his reasoning leads naturally to the claim that all human beings are entitled to the claims of justice:

(4) All beings with the capacity for moral personality are entitled to the claims of justice.
(5) All human beings have the capacity for moral personality.

Therefore,

(6) All human beings are entitled to the claims of justice.

Nevertheless, perhaps this preliminary analysis is too hasty. After all, is Rawls really committed to premise (4)? Is having a capacity for moral personality really enough to be entitled to the claims of justice? And even if it is enough, is it really true that *all* human beings have it (which is what premise (5) claims)? What about those unfortunate human beings who are born into this world with a crippling genetic

[16] Rawls (1971, p. 505). Emphasis [and brackets] mine.
[17] Rawls (1971, p. 505).
[18] Rawls (1971, p. 509).

defect that effectively guarantees that they will never develop the ability to think rationally, deliberate morally, or have a sense of justice? To treat these questions, we need to inspect more closely the remainder of Rawls' general account of the basis of equality.

Immediately after his claim that the capacity for moral personality is a potentiality that is ordinarily realized in due course, Rawls claims that having this potentiality is *sufficient* for being entitled to equal justice, noting that nothing "beyond the essential minimum" is required. He abandons his initial aim of justifying why *all and only* human beings (as opposed to animals) are entitled to the claims of justice, and limits his task to one of justifying why *all* human beings are entitled to the claims of justice. "Whether moral personality is also a necessary condition I shall leave aside."[19] So much for premise (4); we are safe in claiming that Rawls is committed to it. What about premise (5)?

Even the limited task of justifying why *all* human beings are entitled to the claims of justice turns out to be considerably complex for Rawls. In some places, he seems to think that there are some human beings who do *not* possess even the "essential minimum" capacity for moral personality. For example, this thought is lurking in the background even where Rawls' emphasis is on the "overwhelming majority" of human beings who *do* have this capacity:

> ...the capacity for a sense of justice is possessed by the overwhelming majority of mankind.... Even if the capacity were necessary, it would be unwise in practice to withhold justice on this ground. The risk to just institutions would be too great.[20]

These last two sentences are left tantalizingly undeveloped. But presumably they anchor the just treatment of a minority of human beings (those permanently incapacitated, as far as moral personality is concerned), not in the properties of those incapacitated human beings, but in the properties of *other* human beings (those in the "overwhelming majority"). In other words, it is not considerations of *entitlement* that secure the just treatment of those permanently incapacitated, but considerations of *pragmatism*. The idea is that we do not want to even *start* withholding justice from human beings whom we judge not to have the (higher-order) capacity, because to do so inadvertently puts other human beings who really do have the (higher-order) capacity at risk.

However, in other places Rawls seems to think that even the most unfortunate of human beings are merely lacking the *realization* of the higher-order capacity for moral personality, and not the capacity itself. This ambiguity is evident in this sentence:

> Only scattered individuals are without this capacity, or its realization to the minimum degree, and the failure to realize it is the consequence of unjust and impoverished social circumstances, or fortuitous contingencies.[21]

---

[19] Rawls (1971, p. 509).

[20] Rawls (1971, pp. 505–506).

[21] Rawls (1971, p. 506).

Does Rawls think these scattered individuals are truly lacking *the capacity* in question, or does he think they are merely lacking the *realization* of this capacity to the minimum degree? To recall our earlier distinction, the question is whether certain individuals are merely lacking the *immediate* capacity, or whether they are lacking *both* the immediate capacity *and* the higher-order capacity to have this immediate capacity. The following passage shows Rawls is aware of the problem this ambiguity might pose, although he chooses not to fully investigate it:

> A full discussion would take up the various special cases of lack of capacity ... those more or less permanently deprived of moral personality may present a difficulty. I cannot examine this problem here, but I assume that the account of equality would not be materially affected.[22]

Some of Rawls' critics view his decision to avoid a full discussion of this problem (combined with his optimism regarding its eventual solution) as indicative of a serious weakness (or serious blind spot) in his account.[23] But there are two ways to solve this problem, corresponding to the two options for resolving the ambiguity, and on either solution, Rawls can at least get to the idea that all human beings should be treated *as if* they are entitled to the claims of justice.

One way of solving the problem is to develop Rawls' enigmatic remark about the pragmatic foolishness of withholding justice from humans who supposedly lack the relevant capacity ("the risk to just institutions would be too great"). By thinking through the likely consequences of such a policy, we can see why the parties to the original position—who are themselves risk-averse—would reject it. Consider, for example, the fact that for every group of human beings whose members (supposedly) *lack* the relevant capacity, there is a second group of human beings whose members are difficult to distinguish from those in the first group, even though those in the second group *possess* the relevant capacity. Even if the parties to the original position were not concerned about being in the first group, they would be concerned about being in the second group; and since the two groups are difficult to distinguish, the parties would prudentially opt for protecting both groups rather than neither. For another example, consider the way humans have displayed a regrettable tendency to use the category of "human beings undeserving of justice" in the past. Humans with political powers have a tendency to manipulate such a category in order to achieve political goals, with resultant risks to the justice of their institutions. The history of eugenics and human experimentation shows how medical, political, and military institutions can become unjust in this way. Since the parties to the original position know these tendencies, they would have good reasons for avoiding these risks as much as possible.

But a more straightforward way to remove the problem would be to simply develop Rawls' remark about "the failure to realize" a capacity. By attending more carefully to the notion of a hierarchy of capacities, *all* human beings can be seen to

[22] Rawls (1971, pp. 509–510).

[23] See Dombrowski (1997, p. 59). See also Warren (1997, p. 105).

have the morally decisive higher-order capacity in question, even if not all human beings *realize* that capacity.

Recall my earlier example of the person with the chemical imbalance in her brain. That person, while not possessing the immediate capacity to act rationally for her good, nevertheless did possess a higher-order capacity to have this immediate capacity. This higher-order capacity, admittedly, would not be realized unless certain conditions were to be met—namely, the correction of the brain anomaly. But this should not cause us to be skeptical about the *existence* of this higher-order capacity. A human being can have a capacity even if certain physical conditions permanently block its realization. Presumably the same sorts of things would be said about other problematic cases, such as those who suffer life-changing accidents which permanently reduce them to having the mental life of a newborn infant, or those who have inherited genetic traits which guarantee they shall not exercise rational-decision-making unless dramatic genetic therapies enable these genetic flaws to be overcome. For example, human infants with anencephaly actually *do* have the higher-order capacity for moral personality; it is, of course, seriously undeveloped and will likely not be developed without great advances in medical technology. But an undeveloped higher-order capacity is a higher-order capacity all the same.

Hence there are two ways that Rawls could handle the question of how to treat a human being who is purportedly "permanently deprived of moral personality" or "lacking the requisite potentiality". He could admit that there are such human beings, and then flesh out the remark about how withholding justice to these human beings would pose too great a risk to just institutions. Alternatively, he could simply refuse to admit that there are such human beings, since any human being who purportedly fits this description can be more accurately described as in fact *having* the "requisite potentiality," albeit in a seriously undeveloped state. Either way, Rawls is committed to the view that all human beings should be treated *as if* they are entitled to the claims of justice.

Accordingly, there are two ways of formalizing Rawls' *general* account of the basis of equality. The first way is to reject the suggestion that there are scattered individual human beings who lack the requisite potentiality. If we take a more nuanced and consistent view of capacities (as Rawls sometimes seems to), there are no such individual human beings, and the preliminary analysis would stand, arguing from (4) and (5) to (6). The second way of formalizing the account admits the suggestion (as Rawls sometimes seems to) that there are some human beings who lack the requisite potentiality, and so premises (5) and (6) must be amended:

(4) All beings with the capacity for moral personality are entitled to the claims of justice.

(5a) *The overwhelming majority of* human beings have the capacity for moral personality.

Therefore,

(6a) *The overwhelming majority of* human beings are entitled to the claims of justice.

To deliver the conclusion that Rawls is really after, we would need to add some extra premises about the need to safeguard just institutions by protecting even the "scattered individuals" who purportedly lack the capacity in question:

(7) If we want the risk of just institutions not to be too great, then *all* human beings should be treated *as if* they are entitled to the claims of justice.
(8) We want the risk of just institutions not to be too great.

Therefore,

(9) *All* human beings should be treated *as if* they are entitled to the claims of justice.

The conclusion Rawls is really after is (9). Since (6) from the preliminary analysis also entails (9), Rawls gets (9) either way.

The basic thrust of the argument connecting these Rawlsian discussions to the claim this chapter is seeking to defend should be clear. The constraints of justice, and the basic rights corresponding to the constraints of justice, form a significant part of serious moral status. A party to the original position would attempt to guarantee that others in society treated her as if she had this part of serious moral status during the phases of her life when she is relatively incapacitated. What is this part of serious moral status based on? The answer is, the "capacity", at some order or other, "to take part in and to act in accordance with the public understanding of the initial situation." Since these capacities are among the set of typical human capacities, it follows that, if something has a set of typical human capacities, it has at least this part of serious moral status.

7 Capacities and the Capabilities Approach

Martha Nussbaum and Amartya Sen have long been formulating a "capabilities approach" to moral and political philosophy. Although Nussbaum's formulation of the approach is a political view rather than a comprehensive moral view, and although she has developed and changed it in important ways over the years, two features of an intermediate stage of her formulation of the approach are especially relevant to the present chapter.[24] First, if an individual has a certain set of what

[24]The examples of Nussbaum's formulation of the capabilities approach that I will be working with are primarily from an intermediate stage in her development of the approach. They are found in Nussbaum (2000, 1988, pp. 145–184, 1992, pp. 202–246, 1997, pp. 273–300 [but the version I quote from is found on pp. 117–149 of Pablo De Greiff and Ciaran Cronin (eds.), *Global Justice and Transnational Politics: Essays on the Moral and Political Challenges of Globalization* (Cambridge, Mass.: MIT Press, 2002)]). For an example of Sen's formulation of the approach, see his essay (1993). As we shall see below, very major changes in the concept of basic capabilities have occurred in the last several years. For example, in her 2006 book *Frontiers of Justice* (Cambridge: Harvard University Press, 2006) and in her 2008 article "Human Dignity and Political Entitlements", President's Council on Bioethics, 2008), she explains that she has dropped the concept of basic capabilities as originally articulated, because of her concern that it does not make

she calls "basic capabilities", then the state is morally required, as a matter of justice, to provide that individual with the necessary conditions for the development of those capabilities. Second, if an individual has this set of basic capabilities, then the individual has certain natural human rights, such as the right to life. If these two claims of Nussbaum's are true, then it seems that if an individual has this set of basic capabilities, then it has what I have called "serious moral status." But the set of "basic capabilities" Nussbaum focuses on is best understood as a subset of what I have called the "typical human capacities" at some higher-order level. Hence, Nussbaum's capabilities approach seems to imply that if something has a set of typical human capacities at some higher-order level, it has serious moral status.

The argument I wish to defend in this section thus runs as follows:

(1) If something has a certain set *S* of basic capabilities, it has serious moral status.
(2) If something has a set of typical human capacities, it has a certain set *S* of basic capabilities.

Therefore,

(3) If something has a set of typical human capacities, it has serious moral status.

I shall defend this argument by explaining what Nussbaum calls "basic capabilities", by explaining the moral work she once thought such capabilities do, and by responding to possible objections to this interpretation.

Nussbaum's discussions of basic capabilities are found in several of her works. In a 1998 paper she called basic capabilities "B-capabilities" and defined them as follows:

> A person is B-capable of function A if and only if the person has an individual constitution organized so as to A, given the provision of suitable training, time, and other instrumental necessary conditions.[25]

She noted that, in this sense of capability, Aristotle thought that

> ...a boy is capable of functioning as a general (*De An.* 417b30); a myopic person is capable of seeing well (cf. *Metaph.* V.22); an embryo is capable of seeing and hearing; an acorn is capable of becoming a tree...[26]

In a 1997 paper Nussbaum claimed that basic capabilities are

> ...the innate equipment of individuals that is the necessary basis for developing the more advanced capability. Most infants have from birth the *basic capability* for practical reason and imagination, though they cannot exercise such functions without a lot more development and education.[27]

---

enough room for the complete political equality of people with severe mental and cognitive disabilities. In its place, she says that every person born of two parents who possess the capacity for some type of striving and agency is a fundamental equal of every other. She also extends the idea of capability to non-human animals, although she gives the species a special importance. Thanks are due to Nussbaum (personal correspondence) for clarifying to me these changes in her view.

[25] Nussbaum (1988, p. 191).

[26] Nussbaum (1988, p. 191).

[27] Nussbaum (2002, p. 132).

In her 2000 book *Women and Human Development*, Nussbaum claimed that

> These [basic] capabilities are sometimes more or less ready to function: the capability for seeing and hearing is usually like this. More often, however, they are very rudimentary, and cannot be directly converted into functioning. A newborn child has, in this sense, the capability for speech and language, the capability for love and gratitude, the capability for practical reason, the capacity for work.[28]

What Nussbaum in these passages calls a basic capability is the same as what I have called a higher-order capacity. To say that an infant has from birth the basic capability to speak English is just to say that the infant has what I have called a higher-order capacity to obtain the first-order capacity to speak English.

In order to avoid misunderstanding here, it is important to note that what Nussbaum calls a "lower-level" capability is the same as what I have called (following C. D. Broad and others) a "higher-order" capacity. Whereas she would say that an infant has only a lower-level capability to speak English, I would say that it has only a higher-order capacity to speak English. What matters is the reality of a certain sort of nested hierarchy of powers, not whether we label those powers "capabilities" or "capacities", or whether we label the layers of that hierarchy "levels" or "orders", or whether we call a given layer of the hierarchy "higher" or "lower".

Nussbaum claimed in 2000 that basic capabilities are "a ground of moral concern."[29] To understand why, it is important to see how she presented the basic moral intuition behind the capabilities approach, how she contrasted basic capabilities with what she calls *combined* capabilities, and how she made the production of certain combined capabilities the goal of public policy.

"The basic intuition from which the capability approach begins," according to Nussbaum, is that "certain human abilities exert a moral claim that they should be developed":

> Not all actual human abilities exert a moral claim, only the ones that have been evaluated as valuable from an ethical viewpoint. (The capacity for cruelty, for example, does not figure on the list.).... Human beings are creatures such that, provided with the right educational and material support, they can become fully capable of all these human functions. That is, they are creatures with certain lower-level capabilities (which I call "basic capabilities") to perform the functions in question. When these capabilities are deprived of the nourishment that would transform them into the higher-level capabilities that figure on the list, they are fruitless, cut off, in some way but a shadow of themselves. When a turtle is given a life that affords a merely animal level of functioning, we have no indignation, no sense of waste and tragedy. When a human being is given a life that blights powers of human action and expression, that does give us a sense of waste and tragedy...[30]

Let the set of basic capabilities to perform valued functions be named *S*. Every member of *S* is a member of the set of typical human capacities, but not every member of the set of typical human capacities is a member of *S*. For example, the

---

[28]Nussbaum (2000, p. 84). Her text really does say "a capacity for work" rather than "a capability for work".

[29]Nussbaum (2000, p. 84).

[30]Nussbaum (2000, p. 83).

higher-order capacity for cruelty is a member of the set of typical human capacities but not *S*. Nevertheless, if something has a set of typical human capacities, it has a certain set *S* of basic capabilities.

Nussbaum originally elucidated basic or B-capabilities by contrasting them with *internal* or I-capabilities and *external* or E-capabilities,[31] but in subsequent work she calls the external capabilities "combined" capabilities. Internal capabilities are

> ...states of the person herself that are, so far as the person herself is concerned, sufficient conditions for the exercise of the requisite functions. A woman who has not suffered genital mutilation has the *internal capability* for sexual pleasure; most adult human beings everywhere have the *internal capability* to use speech and thought in accordance with their own conscience.[32]

Combined capabilities, on the other hand, are

> ...internal capabilities *combined with* suitable external conditions for the exercise of the function. A woman who is not mutilated but is secluded and forbidden to leave the house has internal but not combined capabilities for sexual expression (and work, and political participation). Citizens of repressive nondemocratic regimes have the internal but not the combined capability to exercise thought and speech in accordance with their conscience.[33]

The production of these combined capabilities, Nussbaum claims, is the goal of public policy. This means "promoting the states of the person by providing the necessary education and care," but it also means "preparing the environment so that it is favorable for the exercise of practical reason and the other major functions."[34]

It is now possible to explain how Nussbaum used to argue that an individual's basic capabilities are morally relevant to public policy in at least two related ways: first, basic capabilities were the basis (the necessary and sufficient conditions) for receiving a just political distribution of the combined capabilities; second, basic capabilities were the basis for natural human rights such as the right to life.

Consider first the relationship between basic capabilities and just political distribution. One of Nussbaum's stated purposes in "Nature, Function, and Capability" was "to describe a promising view about the basis and aims of political distribution,"[35] and part of her summary of this promising view went as follows:

> The aim of political planning is the distribution to the city's individual people of the conditions in which a good human life can be chosen and lived. This distributive task aims at producing capabilities. That is, it aims not simply at the allotment of commodities, but at making people able to function in certain human ways. *A necessary basis for being a recipient of this distribution is that one should already possess some less developed capability to perform the functioning in question.* The task of the city is, then, to effect the transition from one level of capability to another.[36]

---

[31] The section where she explained the differences between basic, internal, and external capabilities was even titled "Levels of Capability." See Nussbaum (1988, p. 186).

[32] Nussbaum (2002, p. 132).

[33] Nussbaum (2002, p. 132).

[34] Nussbaum (2002, p. 132).

[35] Nussbaum (1988, p. 145).

[36] Nussbaum (1988, pp. 145–146), emphasis mine.

Nussbaum claimed that once we recognize that an Aristotelian lawgiver seeks the goal of enabling people to live well and do well, we still need to confront the difficult question of "*for whom* will this goal be sought, and on what basis?"[37] She then set out to argue for the idea that "the *basis* of distribution is a lower-level capability of the person, an untrained natural capability to attain the higher functioning level, given the addition of certain further distributable conditions."[38]

Nussbaum's argument in 1988 for treating this untrained natural capability as a *necessary* condition for being a recipient of developed ethical/intellectual capabilities emerged from Aristotle's discussion of women and natural slaves.[39] She also asked, "Is the presence of a B-capability (or B-capabilities) sufficient as well as necessary for being a subject of the lawgiver's concern? I am inclined to think that it is."[40] Nussbaum gave several reasons for her inclination for treating this untrained natural capability as a *sufficient* condition for being a recipient of developed ethical/intellectual capabilities.[41]

One of these reasons focused on Aristotle's discussion of the morally relevant criteria for the distribution of offices in the city:

> [Aristotle's] argument makes the more general point that in each area, when what we are distributing are the necessary material conditions for a certain function, what we should look to is not irrelevant characteristics (like birth or wealth) but a relevant characteristic, namely, the capability to perform the function in question. Aristotle's example is *aulos*-playing...[42]

Nussbaum then explained why this example is relevant to *basic* capabilities:

> [Aristotle] uses the example to make a more general point: that capability is the morally relevant criterion for distribution of the conditions for a function, since capability (unlike other features) has relevance to the performance of the function. Surely we may apply this general point to the situation in which the legislator is distributing education and other necessary conditions of the I-capabilities. When we do so, we find that the characteristic of persons to which he should look is not birth, or wealth, or good looks, but the presence of a B-capability to perform the function in question.[43]

Finally, Nussbaum made the conceptual link between basic capabilities and justice:

> And the aulos example implies, furthermore, that it is *unjust* if the legislator does not give the auloi to the capable players...Applying this further point to the case of education, we would be entitled to say that it is unjust of the legislator not to give these essential goods to those who are by nature capable of using them.[44]

---

[37] Nussbaum (1988, p. 160).

[38] Nussbaum (1988, p. 160).

[39] Nussbaum (1988, p. 166).

[40] Nussbaum (1988, p. 166).

[41] See Nussbaum (1988, p 166).

[42] Nussbaum (1988, p. 167).

[43] Nussbaum (1988, p. 167).

[44] Nussbaum (1988, p. 192).

Nussbaum summarized the account as follows:

> Aristotle is [saying]. . .to those who have a B-capability, give as much of the relevant goods as would be required to bring that person along from a B-capability to an E-capability. . .[45]

So, then, with these parts of Nussbaum's 1988 discussion of just distribution before us, the following moral principle seems to emerge:

> If something has a certain set *S* of basic capabilities (that are possessed by normal adult human beings, e.g. the basic capability for practical reason and the basic capability for imagination), then justice requires that the legislator give it as much of the essential goods as it needs in order to develop and use these basic capabilities.

And this moral principle, in turn, resonates with and helps to justify the following one:

> If something has a certain set *S* of basic capabilities, it has serious moral status.

This, then, is the relevance of Nussbaum's discussion of the relationship between basic capabilities and distributive justice.

Next, consider Nussbaum's discussion of the relationship between basic capabilities and natural human rights. In 1997 she argued that rights and capabilities are related in two ways. The first relation involved thinking of certain *combined* capabilities as themselves rights:

> The right to political participation, the right to religious free exercise, the freedom of speech, the freedom to seek employment outside the home. . .all of these are best thought of as human capacities to function in ways that we then go on to specify.[46]

The second relation involved thinking of certain *basic* capabilities as the criteria or ground of rights:

> . . .there is another way in which we use the term "right" in which it [sic] could not be identified with a capability. We say, that is, that A has "a right to" seek employment outside the home, even when her circumstances obviously do not secure such a right to her. When we use the term "human right" this way, we are saying that just in virtue of being human, a person has a justified claim to have the capability secured to her: so a right in that sense would be prior to capability and a ground for the securing of a capability. "Human rights" used in this sense lie very close to what I have called "basic capabilities," since typically human rights are thought to derive from some actual feature of human persons, some untrained power in them that demands or calls for support from the world.[47]

Nussbaum's attempt to forge a close connection between basic human capabilities and rights was further clarified by her answer to the question of whether the language of rights is still needed once we have the language of capabilities. She claimed that the language of rights still plays important roles in public discourse:

---

[45] Nussbaum (1988, pp. 192–193).

[46] Nussbaum (2002, p. 135).

[47] Nussbaum (2002, p. 136). As she puts it in *Frontiers of Justice* ". . .the 'basic capabilities' of human beings are sources of moral claims wherever we find them: they exert a moral claim that they should be developed and given a life that is flourishing rather than stunted" (p. 278).

> ...there is no doubt that one might recognize the basic capabilities of people and yet still deny that this entails that they have rights in the sense of justified claims to certain types of treatment...appealing to rights communicates more than appealing to basic capabilities: it says what normative conclusions we draw from the fact of the basic capabilities.[48]

Nussbaum believed that the fact that an individual has certain basic capabilities can appropriately generate normative conclusions about that individual's rights. The idea is that, if an individual possesses a certain set *S* of basic capabilities, it follows that this individual has certain natural rights.

This link between capabilities and rights is especially relevant to serious moral status. For serious moral status, as I argued in Chapter One, includes a strong moral presumption against being killed, and the right to life has traditionally been thought of as one of the natural rights. Indeed, Nussbaum has consistently attempted to produce a working list of "what the most central human capabilities are,"[49] and the very first capability on the current version of her list is life itself:

> 1. *Life*. Being able to live to the end of a human life of normal length; not dying prematurely, or before one's life is so reduced as to be not worth living.[50]

So, then, with these parts of Nussbaum's discussion of human rights before us, the following moral principle seems to emerge:

> If something has a certain set *S* of basic capabilities, it has natural human rights, such as the right to life.

And this moral principle, in turn, resonates with and helps to justify the following one:

> If something has a certain set *S* of basic capabilities, it has serious moral status.

This, then, is the relevance of Nussbaum's discussion of the relationship between basic capabilities and rights.

We can now summarize the argument of this section. If something has a certain set *S* of what Nussbaum calls "basic capabilities", then, according to her, the state is

---

[48]Nussbaum (2002, pp. 138–139). However, notice both the similarity and the shift in *Frontiers of Justice* (Cambridge, Mass.: Harvard University Press, 2006): "the language of capabilities...gives important precision and supplementation to the language of rights....the capabilities approach holds that the basis of a claim is a person's existence as a human being—not just the actual possession of a set of rudimentary 'basic capabilities,' pertinent though these are to the more precise delineation of social obligation, but the very birth of a person into the human community. Thus Sesha [Kittay]'s entitlements are not based solely upon the actual 'basic capabilities' that she has, but on the basic capacities characteristic of the human species. Even if Sesha does not have the capacity for language, then, the political conception is required to arrange vehicles of expression for her, through adequate forms of guardianship. Such entitlements would not exist were capabilities based only on individual endowment, rather than on the species norm" (pp. 284–285).

[49]Nussbaum (2002, p. 128). "The List is supposed to be a focus for political planning, and it is supposed to select those human capabilities that can be convincingly argued to be of central importance in any human life, whatever else the person pursues or chooses" (p. 128).

[50]Nussbaum (2002, p. 129, 2000, p. 78, 2008, p. 377). http://www.bioethics.gov/reports/human_dignity/human_dignity_and_bioethics.pdf

morally required, as a matter of justice, to provide it with the necessary conditions for the development of those capabilities. Furthermore, if something has a set *S* of basic capabilities, then it has certain natural human rights, such as the right to life. But if these things are true, then it also seems true that if something has a certain set *S* of basic capabilities, it has serious moral status. Finally, it can be shown that a set *S* of basic capabilities is merely a subset of the set of typical human capacities. Hence, if something has a set of typical human capacities, at some order or other, it has serious moral status.

There are three related elements in Nussbaum's writing that might seem to undermine the argument of this section. First, she defines "human being" differently than I do. Second, she does not think that all of the entities that I would call human beings have the relevant set of basic capabilities. Third, she sometimes views sentience as a necessary condition of "moral considerability". However, since she does not always view sentience this way, and is currently searching for a solid argument for so viewing it, I shall focus my discussion below only on the first two elements.[51]

For me, the concept of a human being is a purely *descriptive* concept, picking out all and only those entities in the world that have a certain sort of cell structure (the arguments for this were developed in Chapter One); for Nussbaum, the concept of a human being is an irreducible *normative* concept, picking out all and only those entities in the world that exhibit certain functional properties we deem to be morally significant. So, for example, the proposition "this human individual remains a human being during all the stages of his struggle with senile dementia" is simply true on my account, but it is not simply true, and perhaps not even true at all, on Nussbaum's account. As she noted in a 1992 article:

> I have argued that the conception of the human being is itself, in a certain way, a normative conception, in that it involves singling out certain functions as more basic than others. And there is no getting around the fact that correct application of the concept will involve

[51] In *Frontiers of Justice,* Nussbaum admits that there are good reasons for holding that "the capabilities approach...strictly speaking, should not say that the capacity to feel pleasure and pain is a necessary condition of moral status" (p. 362, see discussion on pp. 361–362 and p. 449 fn 54). And in "Human Dignity and Political Entitlement" (President's Council on Bioethics, 2008): "As for when human dignity begins to assert its ethical claims, I have so far argued that sentience is a necessary condition of moral considerability...I have no very solid argument for this position, and I have for some years urged the young members of the Human Development and Capability Association to work out alternative positions on the question, 'Whose capabilities count?'" (pp. 373–374). She there briefly considers how this view about sentience might apply to debates about abortion and embryonic stem-cell research. Since Nussbaum is currently searching for a solid argument in favor of the view that sentience is necessary for moral status, since she is aware of good reasons against that view, and since she is currently urging younger scholars to work out alternative positions to that view, perhaps it will not be thought inappropriate to point out that the arguments advanced earlier in this section, and indeed all the argument advanced earlier in this chapter, explain why sentience should not be thought necessary for moral considerability, and point the way to a better alternative. For example, if a person can temporarily lose her sentience for a period of time without losing her moral considerability during that period of time, then sentience is not a necessary condition for moral considerability, even if the basic capability for sentience (what I call the "higher-order capacity" for sentience) is.

> answering evaluative questions that will sometimes be difficult to answer: for a creature falls under the concept only if it possesses some basic, though perhaps altogether undeveloped, capability to perform the functions in question. It will sometimes be very difficult to say whether a certain patient with senile dementia or a certain extremely damaged infant has enough of those basic capabilities to fall under the concept.[52]

Before explaining how Nussbaum has developed her views regarding these difficult cases, let me simply note that I believe she is incorrect in thinking that certain patients with senile dementia, or certain extremely damaged infants, lack the basic capabilities she wants to focus on. The arguments for this belief have already been given a detailed exposition in Chapter Two of this book. If she wished, Nussbaum could employ those arguments to show how even severely damaged infants and patients with senile dementia are still "human" (even in her normative sense) and endowed with the relevant basic capabilities.

In 1992, Nussbaum had two related strategies for handling these marginal cases. First, she re-emphasized the importance of *basic* capacities in judging whether an individual is "human" (in her normative sense).[53] Second, she admitted that, even though there may be unfortunate cases where our relatives do not have the relevant basic capacities (and are thus not "human" in her normative sense), we should nevertheless proceed *as if* everyone related to us *does* have the relevant basic capacities.[54]

---

[52]Nussbaum (1992, p. 227). For her treatment of this issue in *Frontiers of Justice*, consider: "...the notion of human nature in my theory is explicitly and from the start *evaluative*, and, in particular, *ethically evaluative*: among the many actual features of a characteristic human form of life, we select some that seem so normatively fundamental that a life without any possibility at all of exercising one of them, at any level, is not a fully human life, a life worthy of human dignity, even if the others are present. If enough of them are impossible (as in the case of a person in a persistent vegetative state), we may judge that the life is not a human life at all, any more" (p. 214); "Some types of mental deprivation are so acute that it seems sensible to say that the life there is simply not a human life at all, but a different form of life. Only sentiment leads us to call the person in a persistent vegetative condition, or an anencephalic child, human" (p. 220); "surely it would be only dogmatism to insist that the life of such a child [with anencephaly] is a human life..." (p. 492, fn 25). Her reason for thinking this with the cases of anencephaly and PVS is that "all possibility of conscious awareness and communication with others is absent" (p. 221).

[53]For an example of the first strategy, consider the following passage: "So far, I have focused on the higher-level (developed) human capabilities that make a life a good human life but have not spoken at length about the empirical basis for the application of the concept 'human being' to a creature before us. The basis cannot, of course, be the presence of the higher-level capabilities on my list, for one of the main points of the list is to enable us to say, of some being before us, that this being might possibly come to have these higher-level capabilities but does not now have them. It is that gap between basic (potential) humanness and its full realization that exerts a claim on society and government. What, then, is to be the basis for a determination that this being is one of the human beings, one of the ones whose functioning concerns us? I claim that it is the presence of a lower-level (undeveloped) capability to perform the functions in question, such that with the provision of suitable support and education, the being would be capable of choosing these functions." Nussbaum (1992, pp. 227–228).

[54]For an example of the second strategy, consider this passage: "There is, of course, enormous potential for abuse in determining who has these basic capabilities. The history of IQ testing is

Subsequent work by Nussbaum on this topic has shifted to a third strategy for handling the range of difficult cases. As she described the shift in 2008,[55]

> In early formulations of the idea, I said that the ground of political entitlements lay in a set of "basic capabilities," undeveloped powers of the person that were the basic conditions for living a life worthy of human dignity. I acknowledged that the potential for abuse in assessing which children of human parents have the basic capabilities was very high, and that many groups (women, members of minority races, people with a variety of disabilities) had been prematurely and wrongly said not to have some major basic capabilities (rationality, the capacity for choice, and so forth). So in practical terms I took the line that it was always best to proceed as if everyone was capable of all the major internal capabilities, and to make tireless efforts to bring each one up above the threshold. I still believe that this practical approach is essentially correct. I do think, however, that it is quite crucial not to base the ascription of human dignity on any single "basic capability" (rationality, for example), since this excludes from human dignity many human beings with severe mental disabilities. Even if we should shift to some different capacity, such as the capacity for social interaction or care, many human beings would still be excluded.

After remarking on the complexity of the problem, she suggests that

> ...the best way to solve this complex problem is to say that full and equal human dignity is possessed by any child of human parents who has any of an open-ended disjunction of basic capabilities for major human life-activities.

Finally, to illustrate how she thinks this works out with particular cases, she says:

> At one end, we would not accord equal human dignity to a person in a persistent vegetative state, or an anencephalic child, since it would appear that there is no striving there, no reaching out for functioning. On the other end, we would include a wide range of children and adults with severe mental disabilities, some of whom are capable of love and care but not of reading and writing, some of whom are capable of reading and writing but severely

---

just one chapter in an inglorious saga of prejudiced capability testing that goes back at least to the Noble Lie of Plato's *Republic*. Therefore we should, I think, proceed as if every offspring of two human parents has the basic capabilities, unless and until long experience *with the individual* has convinced us that damage to that individual's condition is so great that it could never in any way, through however great an expenditure of resources, arrive at the higher capability level. (Certain patients with irreversible senile dementia or a permanent vegetative condition would fall into this category, as would certain very severely damaged infants. It would then fall to other moral arguments to decide what treatment we owe to such individuals, who are unable ever to reach the higher capabilities to function humanly. It certainly does not follow that we would be licensed to treat such individuals harshly; we simply would not aim at making them fully capable of the various functions on our list.)" Nussbaum (1992, p. 228). Again, in response, I believe that there is no human individual who "could never in any way, through however great an expenditure of resources, arrive at the higher capability level." The arguments for this belief are given a detailed exposition in Chapter 2. Indeed, I would say of the anencephalic and the patient in a persistent vegetative state exactly what Nussbaum says about Sesha Kittay in the following passage from *Frontiers of Justice*: "if we could cure her condition and bring her up to the capabilities threshold, that is what we would do, because it is good, indeed important, for a human being to be able to function in these ways. If such a treatment should become available, society would be obliged to pay for it, and would not be able to offer the excuse that she is impaired 'by nature.' And, further, if we could engineer the genetic aspects of it in the womb, so that she would not be born with impairments so severe, that, again, is what a decent society would do" (pp. 226–227).

[55]The following passages are all taken from her fuller discussion in Nussbaum (2008, pp. 362–363).

challenged in the area of social interaction. So the notion of "basic capabilities" still does some work in saying why it is so important to give capacities development and expression, but it is refashioned to be flexible and pluralistic, respectful of human diversity.

Two points should be made in response to this these developments within Nussbaum's view. First, there seems to be nothing requiring Nussbaum to abandon her earlier view that the possession of certain basic capabilities is *sufficient* for an entity to have the moral entitlements she thinks normal adult human beings have: for if human dignity is possessed by certain beings who have *even one* of a relevant set of basic capabilities, then surely this dignity is possessed by beings who have the *entire* set of basic capabilities. Second, if the arguments of Chapter 2 are correct, then even the anencephalic and the patient in a persistent vegetative state can be correctly characterized as having the relevant basic capabilities.

I believe a short summary of this chapter would be helpful at this point. The main argument of the chapter was a "moral" temporary change argument. Whereas the "metaphysical" temporary change argument constructed in Chapter 2 attempted to show that being human is sufficient for having a set of typical human capacities, the "moral" temporary change argument of this chapter attempted to show that having a set of typical human capacities is sufficient for having serious moral status. One might naturally ask: why not just skip the detour about typical human capacities, and instead construct a temporary change argument that attempts to show that being human is sufficient for having serious moral status? The answer to this natural question is that focusing upon typical human capacities allows us to explain more: it allows us to explain both why humans retain their serious moral status when undergoing various sorts of temporary changes, and why "shumans" (with capacities just like ours, but with ZNA instead of DNA) retain their serious moral status when undergoing various sorts of temporary changes.

I considered and rejected several alternative strategies for attempting to explain our serious moral status in a way that accommodates our beliefs about temporary changes. Strategies focusing on an individual's actual past and/or future are unsatisfactory, because we can think of cases where an individual still has serious moral status, even though it does not have the sort of actual past and/or future that these strategies recommend. Strategies focusing on a particular order of a given capacity, such as the first-order capacity to think, are unsatisfactory because we can think of cases where an individual still has serious moral status, even though it does not have that particular order of the capacity in question: since the temporary change can always be iterated, there is no principled way to stop as long as there is a capacity *at some order or other* to do the activity in question (e.g. think). Strategies focusing on "actual, continuing subjects of experience" are unsatisfactory, because we can think of cases where an individual still has serious moral status, even though it is not an "actual, continuing subject of experiences" (as that phrase is typically understood by its proponents). The cases I focused on were cases where the "actual, continuing subjects of experience" view is often taken to be most plausible: namely, cases involving the "complete reprogramming" of the upper brain of a human organism.

Finally, I considered why the writings of John Rawls and Martha Nussbaum lend support to the idea that if something has a set of typical human capacities, at some order or other, it has serious moral status. Rawls' own discussions of the way his parties to the original position would deliberate about paternalism and the basis of equality lead naturally to the idea that certain typical human capacities—in particular, certain types of higher-order capacities for moral personality—are sufficient to make something owed justice. Nussbaum's formulation of her capabilities approach lead naturally to the idea that certain typical human capacities—in particular, the "basic capabilities"—are sufficient to make something endowed with certain natural rights and entitled to just treatment by the state.

Chapter 4
Little People: Higher-Order Capacities and the Argument from Potential

If the arguments of Chapters 2 and 3 are sound, then even human infants, fetuses, embryos, and zygotes have the typical human capacities that are sufficient to generate serious moral status. This conclusion was reached by arguing that, as an adult human organism undergoes temporary changes that are more and more serious, the order of the capacities we must appeal to in generating serious moral status gets higher and higher. We eventually reach a point where the adult in the middle of a temporary change has an order of capacities that is just as high as the order of capacities possessed by the most undeveloped human organisms.

But this seems to make the main argument of the book into a version of the Argument from Potential (AFP), which has often been accused of leading to absurd conclusions and of relying on mistakes in moral reasoning. Therefore, in this chapter, I explain why the main argument of the book does not lead to the absurd conclusion that human gametes or somatic cells have serious moral status. I also explain why the main argument does not rely on any of the alleged mistakes in moral reasoning that are made by "potentiality" arguments in ethics.

1 The Dreaded Argument from Potential

The general strategy of the "Argument from Potential," as its name suggests, is to argue *from* the claim that an entity has some potential or other *to* the claim that this entity has some moral status or other. The term "potential" here can mean a variety of things, depending on the philosopher using it, but at the very least this term typically means *mere* potential and thus implies "not actual": for example, a potential "person" is typically not thought to be an actual person, and a potentially rational entity is typically not thought to be an actually rational entity.[1] The AFP argues that

---

[1] Of course, there is nothing in the concept of potential that strictly demands that it be construed as "mere" potential: after all, an entity could be both actually *P* and potentially *P* at the same time. Indeed, as I argued in Chapter 2 with David Hume and David Robinson, the fact that something is actually P usually is good evidence for the idea that it is potentially P: if you ask me whether I have the potential to lift a one hundred pound barbell over my head, it usually settles the matter for me to reply "yes, of course I do—in fact, I'm lifting such a barbell over my head right now."

R. DiSilvestro, *Human Capacities and Moral Status*, Philosophy and Medicine 108,
DOI 10.1007/978-90-481-8537-5_4,

the mere fact that an entity has a certain potential generates certain obligations on the part of others to treat the entity in certain ways, whether or not that potential is now being actualized or has ever been actualized before.

Although this is the general strategy of the AFP, particular versions of it vary. At least three main things account for the differences between the versions. First, as just mentioned, different versions of the AFP employ different meanings of the term "potential". This term can mean one or more of the following: bare logical possibility; probability or likelihood; an active causal propensity to change in certain ways; a passive receptivity to undergo certain sorts of changes.

Second, different versions of the AFP rely on different understandings of what the (given notion of) potential is said to be a potential *for*. Some focus on a given attribute: the potential for consciousness or sentience or rationality. Others focus on the potential for being a certain kind of thing: a potential "person" or a potential human being or a potential right-holder or a potential bearer of interests.

Third, a given version of the AFP is partly a function of just what moral relevance the (given notion of) potential is said to have. The consequentialist versions typically focus on the value of what a given entity's potential is a potential *for*, and thus typically argue that any obligation to "respect" an entity's potential grows out of a more basic obligation to promote certain valuable outcomes. The nonconsequentialist versions typically focus on the moral standing of the entity that possesses the potential, and thus typically argue that certain sorts of potential give a being certain sorts of interests (such as the interest in realizing its potential) or rights (such as a right to life).

The AFP is both important and controversial. The AFP is important because of its history and centrality in philosophical debates about the morality of abortion. Michael Tooley has even claimed that when the conservative position on abortion "is thought through in a critical fashion, it appears to be the case that it [the conservative position] stands or falls with the answer to be given to the question of the moral status of potential persons."[2] He reaches this claim by arguing that the moral status of the fetus is the central issue in the abortion debate[3] and that attempts to defend the moral status of the fetus without relying on the concept of a potential person do not succeed.[4] Similarly, Jim Stone has claimed that "a strong fetal claim to protection rises or falls with the appeal to the fetus's potentiality, for nothing else can justify it."[5]

---

[2]Tooley (1983, p. 169).

[3]For example, in Tooley (1983, Section 3, pp. 40–49), Tooley argues against the idea that a woman's right to control what goes on in her own body (an idea reflected, for instance, in Judith Jarvis Thomson's argument from unplugging the violinist) is the central moral issue in the abortion debate.

[4]For example, in Tooley (1983, Section 4, pp. 50–86), Tooley argues against the moral relevance of species membership.

[5]Stone (1987, p. 815).

Although the AFP is also important to the moral evaluation of infanticide, contraception, and the use of human embryos in medical experiments, it originally emerged out of debates about abortion. This may explain, at least in part, why the AFP is so controversial: in particular, it is both widely criticized and widely loathed. As Michael Wreen once put it, "Potentiality is taking a bit of a beating in ethical circles these days, and the concept, it would be no exaggeration to say, currently enjoys about as much popularity as leprosy or the bubonic plague."[6] In addition to the fact that a variety of argumentative assaults have been carried out against the AFP, there is also the fact that the AFP can trigger some of the most visceral emotional responses among philosophers. As Elizabeth Harman recently put it, "I used to be naively terrified of acknowledging any moral significance for potentiality."[7] Now, lest these quotes mislead, it is worth noting that Wreen ends up arguing that potentiality is not vulnerable to most of the common assaults upon it, and Harman ends up arguing that her terror was premature. Indeed, her quote comes in her paper's very last section, titled "How I Learned to Stop Worrying and Love Potentiality."[8] Although Harman's eventual endorsement of the qualified relevance of potentiality exemplifies the open-mindedness of some philosophers, these quotes still illustrate that a defender of the AFP has her work cut out for her.

It is not the purpose of this chapter to explain and evaluate all of the available versions of the AFP that have emerged in the considerable philosophical literature on this topic in the past 30 years. Rather, the purpose of this chapter is to explain how the main argument of this book, which connects a thing's higher-order capacities to its serious moral status, intersects with this literature. Since I have already argued, in Chapters 2 and 3, that it makes sense to think that undeveloped human organisms possess the capacities that generate serious moral status, the next two sections of this chapter will focus on showing how one can agree with the arguments of Chapters 2 and 3 *without* endorsing certain absurdities that the AFP is alleged to commit one to, and *without* embracing certain fallacies that the AFP is alleged to commit. In particular, Section 2 explains why the present account is not committed to the idea that human precursors like gametes or body cells have serious moral status. Since not all "precursors" to undeveloped human organisms have the same set of typical human capacities as you, it is an open question whether these precursors have serious moral status.[9] Section 3 explains why the present account does not commit the logical error of conflating the mere *potential* qualification for a moral status property (such as rights) with the *actual* possession of such a qualification.

---

[6] Wreen (1986, p. 137).

[7] Harman (2003, p. 194).

[8] Harman (2003, p. 193).

[9] The reason for saying *not all* precursors rather than *no* precursors is that, as we shall see below, in cases of embryonic fission and fusion, some precursors to undeveloped human organisms are themselves undeveloped human organisms.

2 Not Every Cell Is Sacred

The AFP is often accused of leading to the absurd conclusion that human gametes—whether sperm, or egg, or some combination of sperm and egg prior to conception—possess the sort of properties that generate serious moral status. A similar allegation has been made against the AFP regarding human somatic cells—whether the clonable somatic cells of an adult human being, or the "totipotent" somatic cells of a developing human embryo.

I shall summarize the argument of this section in this paragraph and expand upon it in the paragraphs which follow. The main argument of this book does not lead to the absurd conclusion that human gametes or somatic cells have serious moral status, because human gametes and somatic cells do not have a set of typical human capacities. Remember, the set of typical human capacities that I claim are sufficient for serious moral status are "identity-preserving" capacities, not "compositional" capacities. A lump of bronze has the capacity to constitute a statue. But it does not have the identity-preserving capacity to do the things that the statue will do (e.g., cause statue lovers to visualize Socrates). The material bits and particles that gradually enter your lungs and bloodstream, and that eventually migrate to your legs and brain, do have the capacity to constitute, or at least partially constitute, something that walks and feels pain. But these material bits and particles do not have the identity-preserving capacities to walk and feel pain. Likewise, human gametes and somatic cells have the capacity to constitute something that will eventually think. But these precursor cells do not have the identity-preserving capacity to think.

To begin expanding upon this line of reasoning, consider the following concern expressed by Michael Tooley:

> To characterize potential persons as entities that have a passive potentiality for becoming persons would have the consequence that random collections of matter that could, with sufficient knowledge and technological advances, be transformed into human organisms, would have to be classified as potential persons.[10]

Although Tooley's concept of a person is different than mine (see Chapter 1), his concern here can be reformulated in a way that still appears threatening to my project, by replacing the language of potential persons with the language of the higher-order capacity to think:

> To characterize those with a higher-order capacity to think as entities that have a passive higher-order capacity to think would have the consequence that random collections of matter that could, with sufficient knowledge and technological advances, be transformed into thinking things, would have to be classified as entities with the higher-order capacity to think.

This concern amounts to the claim that, by including a thing's passive capacities in its set of higher-order capacities, the result is an unwelcome proliferation, and indeed, explosion, of capacities in the world. Virtually everything will turn out to

---

[10]Tooley (1983, p. 168).

have the higher-order capacity to do everything. For example, random collections of matter will turn out to have the higher-order capacity to think.

If what it means for an entity to "become" a person is understood along the same lines as what it means for a lump of bronze to "become" a statue, then Tooley's quote, and its reformulation in terms of higher-order capacities, is surely correct, although the clause about "sufficient knowledge and technological advances" is not necessary. For the relevant transformations already occur every day without much technology and without much knowledge: all it takes is a living human organism and a certain amount of time. Human organisms are constantly getting transformed, acquiring new parts through processes like inhalation, nutrition, and hydration, and losing current parts through processes like exhalation, excretion, and perspiration. These mereological transformations (from the Greek *meros*, "part") are ubiquitous; they are the very stuff of life itself.

Consider an example of a mereological "switch" in which a human organism and an island of material stuff gradually exchange all their simplest parts with one another between times t and t*. Label the simplest parts constituting the human organism at t the *ps*, and the simplest parts constituting the island at t the *qs*. Label the human organism constituted by the *ps* at t Lew, the island constituted by the *qs* at t Honolulu, the human organism constituted by the *qs* at t* Kareem, and the island constituted by the *ps* at t* Caribbea. Imagine that the rate of exchange among *ps* and *qs* was such that whenever a *q* leaves Honolulu and becomes a part of Lew, a *p* leaves Lew to take the place of that *q*. This example can be diagrammed by letting solid black lines represent the flow of *ps* and *qs* and by letting dotted lines represent the "continuity" of the island and the man:

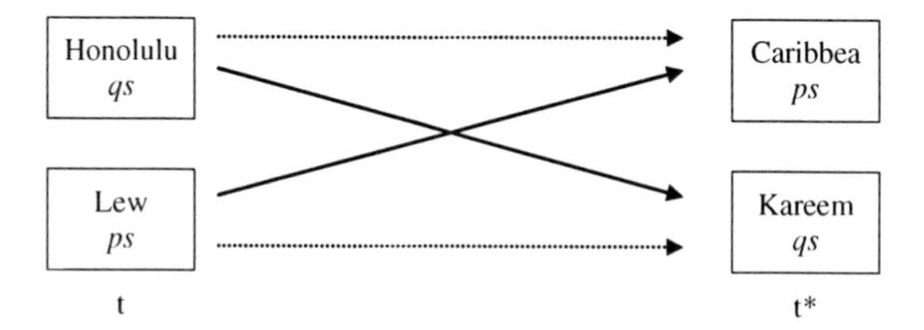

This diagram illustrates several distinct sorts of potentiality:

(1) Honolulu is potentially Kareem.
(2) Lew is potentially Caribbea.
(3) Honolulu is potentially Caribbea.
(4) Lew is potentially Kareem.

The sense of potentiality found in (1) and (2) is what was earlier labeled compositional potency. These claims rely on the constitutive sense of "is": (1) must be paraphrased as, and means no more and no less than, the claim that the *qs* that constitute Honolulu have the potential to constitute Kareem. The same thing is true of (2).

Assume that Kareem has the first-order capacity to speak Chinese, but that Lew does not. If the concept of a higher-order capacity is explicated using the sense of potentiality found in (1) and (2), then Honolulu has the higher-order capacity to speak Chinese. But this is absurd: an island of material stuff surely does not have the higher-order capacity to speak Chinese. This suggests that the concept of a higher-order capacity should not be explicated using the sense of potentiality found in (1) and (2).

The sense of potentiality found in (4) is not compositional potency but what was earlier labeled identity-preserving potency. This claim relies on the "is" of identity: (4) must be paraphrased as, and means no more and no less than, the claim that Lew has the potential to become Kareem while remaining the self-same individual over time.

The concept of a higher-order capacity should be explicated using the sense of potentiality found in (4). Return again to the assumption that that Kareem has the first-order capacity to speak Chinese, but that Lew does not. If the concept of a higher-order capacity is explicated using the sense of potentiality found in (4), then Lew does, while Honolulu does not, have the higher-order capacity to speak Chinese. For Lew does, but Honolulu does not, have the potential to become Kareem while remaining the self-same individual over time.

What about claim (3)? Claim (3) is controversial because philosophers do not agree on what it takes for an island of material stuff to persist through time.[11] If (3) is interpreted like I have interpreted (4), then many will claim (3) is false. For it is not true that Honolulu has the potential to become Caribbea while remaining the self-same individual over time. Islands are not individuals in the same way that organisms are, and islands do not persist in the same way that organisms do. If (3) is to be true at all, it must be paraphrased as, and mean no more and no less than, the claim that Honolulu was able to "become" Caribbea by a gradual replacement of its parts. Of course, if that is all (3) means, then the same thing could be said of (4): for Lew is able to "become" Kareem by a gradual replacement of its parts. Both human organisms and islands have higher-order capacities to be constituted by new sets of particles. But even if I agree with this point, the higher-order capacity to be constituted by new sets of particles is *not* the higher-order capacity I have in mind with (4).

The metaphysical picture I am advocating here can be clarified with more simple examples. Even if the saying is true that you are what you eat, still, you are not what you eat until you eat it, and even then, what you eat is not literally you—it merely composes you: even while it composes you, it does not do all the things you do, and before you ate it, it did not have the higher-order capacities to do all the things you do. Even if God made Adam out of a pile of dust we name Dusty, still, Dusty

[11] Rea (1997).

did not have the higher-order capacities to do all the things Adam did. Even if the saying were true that little boys are made of snips and snails and puppy dog tails, still, snips and snails and puppy dog tails do not have the higher-order capacities to do all the things little boys do (and the same thing goes for little girls, who are supposedly made out of "sugar and spice and all things nice").

With this metaphysical picture in mind, then, we are in a good position to evaluate whether the argument of Chapters 1 through 3 entails that things like human gametes have serious moral status.

If something's serious moral status were grounded in the mere fact that it is a member of a causal sequence which can contribute to producing certain sorts of valuable outcomes, or certain sorts of valuable entities, then human gametes would indeed have serious moral status. For many human gametes are indeed members of such causal sequences. A similar thing could be said about certain human somatic cells. But this book has argued that something's possessing a set of typical human capacities is what grounds its moral status, quite apart from what causal sequences the thing is a part of. Hence, in order to show that the argument of this book entails the claim that there is a strong moral presumption against killing human gametes, or human somatic cells, one needs to show that these gametes, or somatic cells, possess the relevant capacities.

The most direct way of showing this would be to show that the gametes, or the somatic cells, are *identical* to a thing that possesses the relevant higher-order capacities. Because of this, and because Chapter 2 argued that a human zygote possesses the relevant higher-order capacities, it will now be argued that no human gametes (whether considered separately or considered jointly) and no human somatic cells (whether clonable or totipotent) are identical to any human zygote. The basic claim behind this argument is that, whenever we are confronted with a situation where a zygote appears to be identical to one of its "precursors", the situation is better understood as one where all the material that used to constitute the precursors now constitutes the zygote, even though the precursor is not identical to the zygote.

Few would claim that a sperm cell is identical to the zygote it helps to produce. Perhaps the reason for this is due to simple matters of size. Imagine that a statue was composed of 99% bronze and 1% silver. It is hard to believe that the silver that went into the statue is identical to the statue itself. Perhaps considerations of these sorts are what lead us to resist the idea that a sperm cell is identical to the zygote it helps to produce.

However, more must be going on than considerations of relative size in order to deny the claim that an egg cell is identical to the zygote it helps to produce. After all, the overwhelming bulk of the material constituting the zygote constituted the egg a short time earlier. This biological fact makes it seem easier to refer to the zygote as a "fertilized egg" rather than an "enveloped sperm". Indeed, one occasionally hears the suggestion that, because the zygote is referred to as a "fertilized egg", the zygote literally *is* an egg. Perhaps these facts about biology and our linguistic conventions are responsible for way philosophers invoke the individual egg cell as a reason for rejecting the moral relevance of an entity's "potential". For example, when Mary Anne Warren considers John Rawls' claim that the "capacities" that underwrite human equality must be understood in a way that coheres with our intuitions regarding human infants, she claims that his claim leads to absurd results:

> The hypothesis that the potential to develop one's own ends and a sense of justice is sufficient for moral personhood enables Rawls to gather normal infants and young children into the fold. However, it appears to do this at the cost of also admitting fertilized *or unfertilized* human ova—which also have the potential, under the right circumstances, eventually to develop the capacities in question.[12]

Although Warren may be right in suggesting that Rawls' account admits fertilized human ova into the fold, her suggestion that it brings *unfertilized* human ova into the fold relies on the assumption that the unfertilized ova is identical to the entity that develops the capacities in question. But this assumption is mistaken, since the zygote is not identical to the egg. Just as a statue whose composition is 99% bronze and 1% silver is not for that reason identical to the lump of bronze that went into the sculptor's shop, so too a zygote, whose composition consists of materials, most of which came from an egg, is not for that reason identical to the egg. It is not merely considerations of relative size that prevent us from identifying one entity *a* with another entity *b*, but considerations of the difference between *a* continuing to exist as *b* and *a* contributing all of its material to *b*.

The idea that the zygote is not identical to either sperm or egg is confirmed by attempting to construct a temporary change for the sake of gametes that parallels the temporary change constructed above for the sake of zygotes. Recall the modification to Jeff McMahan's thought experiment, where your body can shrink down to a zygote and then grow back into a fully formed adult organism. Now imagine that, instead of making the zygote the turn-around point in this biological cycle, we allow the zygote to split apart into a sperm and an egg, and then we allow the sperm and egg to come back together again and re-fuse, and then we allow the zygote to grow up again, so that the adult organism after the shrinking-fissioning-fusioning-growing process has the same personality traits, etc. as the adult organism before this shrinking-fissioning-fusioning-growing process. Now ask: do you think that either the sperm or the egg in the middle of this process was really you? I believe that most people, when confronted with this prospect, would deny that they were either the sperm or the egg in the middle of this process. Consequently, when confronted with the apparent fact that "you" seem to exist at the end of this process, one must either reject this apparent fact (since individuals cannot have temporal gaps) or else explain how "you" can continue to exist during the fissioning-fusioning part of the process without existing as either the sperm or the egg.

Some of the same things need to be said about parthenogenesis. Parthenogenesis (from the Greek words meaning "virgin birth") takes places when an ovum, without the actual fertilization of a sperm, nevertheless begins to start dividing in the way a zygote divides. Parthenogenesis is thought to be rather rare for humans; although it can be triggered in a laboratory by what biologists call "noxious stimulation," the process of dividing has never been observed to continue for more than a few

[12] Warren (1997, p. 105) emphasis mine.

days.[13] But in some organisms parthenogenesis is the standard way of reproduction. Whether in other organisms or in humans, however, the relevant point is the same: parthenogenetic development should not lead us to assume that an individual egg cell that exists before parthenogenesis is strictly identical to an individual cell that exists after parthenogenesis. Just as a lump of bronze is not identical to the bronze statue that it gives rise to, even if that statue is composed of all and only the bronze that composed the lump, so too an egg is not identical to the zygote that it gives rise to via parthenogenesis, even if that zygote is composed of all and only the material that composed the egg. This is an "in principle" approach to parthenogenesis, which would apply whether or not the number of actual ("in fact") cases of parthenogenesis with human oocytes is zero, one, or quite a large number. Even if embryologists are fully convinced that what they have observed is the development of an embryonic human organism, this "in principle" approach still stands.

The last few paragraphs do not show that there is something incoherent about the claim that the sperm (or the egg) possesses the set of higher-order capacities that generates serious moral status. They merely show that this claim is not *entailed* by the account of higher-order capacities given in Chapters 1 through 3. For the account of higher-order capacities given there does not merely rely upon whether the *time* it takes for a given entity *a* to "give rise to" a given entity *b* is equivalent to the *time* it takes for you to emerge from a certain sort of temporary change. In addition, the account of higher-order capacities given there relies upon whether *a* and *b* are *identical* to one another.

Even if neither the sperm nor the egg, considered separately, has the relevant higher-order capacities, it might be thought that the sperm and egg, considered jointly, do have the relevant higher-order capacities. For it might be thought that

(1) The sperm and egg, considered jointly, do indeed constitute some sort of entity or thing.

and that

(2) This entity or thing is identical to the zygote.

At this point it should come as no surprise that the present account, even if it agreed to (1), is going to disagree with (2) for the reasons given a moment ago: just because a zygote is composed of all and only the material that composed a certain entity, that does not mean that the zygote is identical to that entity. But is there any reason for thinking that (1) is true? Alistair Norcross once flirted with denying the claim that "a combination of sperm and ovum, understood as a mereological sum, is not a thing"[14]:

> Perhaps the most obvious answer to the charge that a combination of sperm and ovum is not a thing is simply to deny it. I am inclined to pursue this option. To the extent that I am

[13] In this paragraph I am relying on David Oderberg for details about parthenogenesis. See Oderberg (1997), especially pp. 282–292.

[14] Norcross (1990, p. 272).

prepared to admit that a zygote or a fetus is a thing, I would claim that a combination of sperm and ovum is also a thing.[15]

But this denial is tantamount to putting mereological sums on the same metaphysical footing as genuine wholes. For example, it puts the mereological sum consisting of your various body organs (or cells, or chemicals, or whatever part-type we choose to focus on with our mereological microscopes) on the same metaphysical footing as you. Indeed, since mereological sums are no respecter of the separateness of persons, Norcross' denial even puts you on the same metaphysical footing as a mereological sum consisting of *your* heart, *my* lungs, and so on. Because we have good reasons for denying that mereological sums in general are on the same metaphysical footing as genuine wholes, and because Norcross offers no reason for his denial apart from his own inclination, it seems that we are still justified in resisting (1).

To sum up this discussion of gametes: one of the reasons the gametes are not metaphysically on a par with the zygote is that neither gamete can claim to be strictly identical with the zygote. Although the gametes, considered jointly before fertilization, have the potential to produce a zygote, and although the gametes are indeed spatiotemporally continuous with the zygote, it is not true that either gamete is the same organism as the zygote. The difference between killing the gametes before fertilization has occurred and killing the zygote after fertilization has occurred is the same difference as that between preventing a human organism from beginning to live and preventing a human organism from continuing to live.

Now exactly the same sorts of considerations are going to be relevant to a proper evaluation of the metaphysics of clonable adult cells and totipotent embryonic cells. For example, in cloning, the living materials that go into a process of somatic cell nuclear transfer are not identical to the living entity that comes out of that process. A temporary change argument can support this judgment. The idea that the entity created by somatic cell nuclear transfer is not identical to either the somatic cell from whence some of its parts came, or to the enucleated egg from which some of its parts came, or to the "mereological fusion" of the somatic cell and enucleated egg, is confirmed by attempting to construct a temporary change for the sake of the somatic cell (and/or the enucleated egg) that parallels the temporary change constructed above for the sake of zygotes. First, recall the modification to Jeff McMahan's thought experiment, where your body can shrink down to a zygote and then grow back into a fully formed adult organism. Next, imagine that, instead of making the zygote the turn-around point in this biological cycle, we allow the zygote to split apart into a somatic cell's extracted nucleus and an enucleated egg, and then we allow the somatic cell's extracted nucleus and the enucleated egg to come back together again and re-fuse, and then we allow the zygote to grow up again, so that the adult organism after the shrinking-fissioning-fusioning-growing process has the same personality traits, etc. as the adult organism before

[15] Norcross (1990, p. 272).

this shrinking-fissioning-fusioning-growing process. Now ask: do you think that either the somatic cell's extracted nucleus or the enucleated egg in the middle of this process was really you? I believe that most people, when confronted with this prospect, would deny that they were either the somatic cell's extracted nucleus or the enucleated egg in the middle of this process. Consequently, when confronted with the apparent fact that "you" seem to exist at the end of this process, one must either reject this apparent fact (since individuals cannot have temporal gaps) or else explain how "you" can continue to exist during the fissioning-fusioning part of the process without existing as either the somatic cell's extracted nucleus or the enucleated egg.

Although such lines of reasoning are persuasive to some, I have noticed that the concern about cloning is one that is particularly hard to shake. Even some people, who have little trouble distinguishing a human zygote from the sperm and egg that preceded it, sometimes have more than a little trouble distinguishing a one-celled human organism produced via SCNT from the somatic cell's extracted nucleus and the enucleated egg that preceded it. Therefore, in the remaining parts of this section, I would like to address in detail one contemporary bioethicist who has claimed that arguments which place much importance on the "potential" (what I call higher-order capacities) of human embryos really do run into a problem with cloning, or somatic cell nuclear transfer.[16]

In a recent issue of the journal *Bioethics*, Massimo Reichlin presents a clarification and defense of the "argument from potential" (AFP) to the conclusion that the human embryo should be protected from the moment of conception.[17] But in her recent essay entitled "Every Cell is Sacred: Logical Consequences of the Argument from Potential in the Age of Cloning" R. Charo claims that versions of the AFP like Reichlin's are vulnerable to a rather embarrassing problem: with the advent of human cloning, such versions of the AFP entail that every somatic cell in the human body ought to be protected.[18] Since this entailment is clearly absurd, she argues, these versions of the AFP should be rejected.

Although Charo is not the only writer who has made this claim,[19] her essay is as good a place as any to begin evaluating it. I shall be arguing that Charo has failed to make her case. There is a relevant distinction between the sort of potential possessed by the somatic cell and the sort of potential possessed by the early human embryo. Since only the latter sort of potential falls within the scope of the AFP, the alleged absurd entailment is no entailment at all. Hence the AFP cannot be rejected on the grounds Charo advances.

---

[16]What follows is adapted from DiSilvestro (June 2006).

[17]Reichlin (January 1997, pp. 1–23).

[18]Charo (2001).

[19]McInerney (1990) and Peters (2001, pp. 127–128).

Charo's central challenge and her pivotal question. The title of Charo's essay trades on the lyrics of the song "Every Sperm is Sacred" from Monty Python's *The Meaning of Life*:

> Every sperm is sacred.
> Every sperm is great.
> If a sperm is wasted,
> God gets quite irate.[20]

Charo thinks this song, which attempts to poke fun at certain forms of religious opposition to contraception, also implicitly pokes fun at the idea that "any entity which has the potential to become a baby must be treated as sacredly as babies", an idea she labels "the classic argument from potential." Such an argument, in an age of cloning, leads to the absurd conclusion that "not only is every embryo or fertilized egg to be treated as sacred, but so should every nucleus-bearing cell in our bodies."[21]

The structure of Charo's essay is twofold. First, she thinks that exploring "the way in which the argument from potential interacts with the new insights gained from cloning technology" provides insights which weaken " the analytical aspects of" and undermine " the logical integrity of" the AFP. Second, she thinks that when we articulate better "the real reasons" for opposition to human embryo research, we will be led to a compromise which allows some destructive embryo research "while at the same time demonstrating some degree of respect and deference to those troubled by cavalier destruction of the possible future children represented by those embryos."[22] I will limit my comments to the first stage of her essay, since it embodies the heart of her cloning-based critique of the AFP.[23]

When Charo describes somatic cell nuclear transfer cloning, she applies to genetic material the distinction between a reversible coma and a permanent vegetative state (PVS). Prior to the 1980s, she claims, we thought "the process of specialization placed all genes in a state akin to a permanent vegetative state (PVS), except those needed to code for the specialized function." But then we discovered that:

> ...by starving the cell into a state amenable to reproduction, and then combining it with the cytoplasm of an enucleated egg cell, the genes previously thought to be in PVS could be awakened and coaxed into performing as if the cell were an ordinary fertilized egg, that

---

[20] *Every Sperm is Sacred.* Lyrics by M. Palin and T. Jones; composed by D. Howman and A. Jacquemin.

[21] Charo (2001, p. 82).

[22] Charo (2001, pp. 83, 87).

[23] Yet even this first stage of her essay contains arguments that cannot be fully addressed here: for example, she argues that certain oft-invoked criteria for moral status (e.g. genetic uniqueness) do not cleanly entail that each early embryo has moral status, since these criteria run afoul of certain biological phenomena (e.g. the possibility that an early embryo can undergo fission (to form twins) or fusion (to form mosaics)).

is, to divide and develop into a fetus, baby, and thence adult sheep. In other words, the previously dormant genes were not in PVS, merely in a reversible coma.[24]

Given this prospect of "using nuclear transfer cloning in humans to actualize the potential of every nucleus-bearing differentiated somatic cell", Charo raises the central challenge of the paper, which I shall label Q1:

[Q1] What distinction is there between ordinary cells and those entities, such as fertilized eggs or embryos, for which the argument from potential is used to justify deferential treatment?[25]

As a way of surfacing one possible answer to Q1,[26] Charo interacts with Massimo Reichlin's essay in order to get the following idea: Reichlin's strategies for answering certain objections to the AFP ultimately rely on viewing the embryo as having a *particular sort* of potential which (a) morally matters and (b) is not possessed by gametes. But Charo then attempts to show that this *same sort* of potential is also possessed by somatic cells, as a way of forcing Reichlin (and others like him) to recognize that somatic cells morally matter in precisely the same sense that embryos morally matter. In particular, Charo notices Reichlin's use of the distinction between *active* potential and *passive* potential in defending (b): the embryo possesses an *active* potential that the gametes lack.[27] While passive potential is "that which would permit the being to develop if subject to external actions that help bring out certain traits that are present, albeit not central to its essence",[28] *active* potentials are "those inherent to the very nature of the being... a tendency which is dependent on its very nature".[29] Charo claims that if the provision of an artificial culture medium to an extra-corporeal fertilized egg "is considered a form of external assistance akin to that at issue in passive potentiality," then the extra-corporeal fertilized egg is *not* a potential baby (since it will not become a baby without the medium). She notes that since opponents of embryo research do not accept a difference between *in vitro* and *in vivo* embryos, they must believe that "whatever external assistance is needed to ensure development of in vitro embryos into babies is consistent with the type of assistance encompassed by the idea of 'active' potential".[30]

This discussion of *active* potential and *assistance* leads directly to Charo posing the following pivotal question, which I will label Q2:

---

[24]Charo (2001, p. 83).

[25]Charo (2001, p. 83).

[26]Charo critiques other possible answers to Q1 as well.

[27]However, Charo does not seem to notice that Reichlin does not think the gametes have *either* passive *or* active potential, but "rather the possibility that two entities unite in order to form a new individual which is distinct from the two originals" (Reichlin, January 1997, p. 4).

[28]Charo (2001, p. 85).

[29]Reichlin (January 1997, p. 14).

[30]Charo (2001, p. 86).

> [Q2] [W]hy [is] the assistance needed to get a skin cell to generate a baby. . .any different from the assistance needed to get an in vitro embryo to develop into a baby [?][31]

She is skeptical that a relevant difference exists, and her reasoning here is well worth mulling over:

> Both [skin cells and in vitro embryos] need a culture medium. For skin cells, that medium is found partly in the cytoplasm of an enucleated egg, but a medium it is nonetheless. The skin cell needs an electric shock. Again, however, it is unclear why electricity, as opposed to the warmth of the incubator used in ordinary management of an in vitro embryo, is of ontological significance.[32]

Note how Q2 is relevant to Q1: if there is no satisfactory answer to Q2, then Q1 cannot be answered merely by appealing to the "active" potential of the embryo.

Answering the pivotal question. Charo is quite right to ask Q2. But in asking it the way she does, she puts her finger on the answer without noticing it. Admittedly, it is not the distinction between *kinds* of assistance that marks the difference between cloning and coaxing an in vitro embryo to develop. Thus, for example, the relevant distinction is not between different *kinds* of culture medium (cytoplasm versus agar) or between different *kinds* of electrical energy input (jolt versus warmth).

What marks the relevant difference is not the distinction between the *kinds* of assistance, but rather the distinction between the *effects* of assistance. Notice the words Charo uses in asking Q2: the effect of one kind of assistance is "to get a skin cell to *generate* a baby", while the effect of the other kind of assistance is "to get an in vitro embryo to *develop into* a baby" (emphasis mine). Now to be sure, Charo's choice of terms is not intended to be anything more than stylistic. Nevertheless, these terms correspond to a genuine substantive distinction between generation and development, and this substantive distinction is relevant to Q2. For x to *develop into* y, it must be true both (1) that x is the same *numerical* thing as y, and (2) that x is the same *kind of* thing as y. For x to *generate* y, however, neither (1) nor (2) are necessary.[33]

To see this distinction at work outside of the context of in vitro embryos and skin cells, consider the difference between a child (x) developing into an adult (y) and a pair of gametes (x) generating a zygote (y).[34] When a child develops into an adult, the child is numerically the same thing as the adult, and is the same kind of thing as the adult—namely, an individual human organism. But when gametes generate a zygote, the gametes are not numerically the same thing as the zygote, nor are they

[31] Charo (2001, p. 86).

[32] Charo (2001, p. 86).

[33] Some might object here that the phrase "generate" sometimes *does* indicate something of one kind producing something else of the same kind. While there may be a sense in which "generate" has such a meaning—namely, that of "begotten, not made" (as in the sentence "did these cats *generate* this entire litter of kittens?"), it is also true that the more common meaning of "generate" is much more neutral—namely, that of "made, not begotten" (as in the sentence "can this machine *generate* any electricity?").

[34] I realize there are complications with referring to a pair of entities with a singular variable; let the variable apply to the gametes "considered jointly".

the same kind of thing as the zygote. The zygote is an individual human organism even though neither gamete is an individual human organism.

Likewise, when an in vitro embryo (x) develops into a baby (y), the embryo is numerically the same thing as the baby, and is the same kind of thing as the baby—namely, an individual human organism. But when a skin cell (x) generates a baby (y), the skin cell is not numerically the same thing as the baby, nor is it the same kind of thing as the baby. The baby is an individual human organism even though the skin cell is not.

The distinction between development and generation corresponds neatly to the distinction between what Stephen Buckle calls the potential to *become* and the potential to *produce*. The potential x has to become y presupposes identity between x and y, while the potential x has to produce y does not presuppose this identity. The zygote has the potential to become a baby, but the skin cell has merely the potential to produce a baby.[35] At this point, two related objections—one for each of the two necessary conditions for "development"—can be introduced. First, it is not obvious why the individual skin cell that generates a zygote is a numerically different thing from the zygote. Second, it is not obvious why the individual skin cell that generates a zygote is a different kind of thing from the zygote.

The first objection can point out that the skin cell before the cloning, and the zygotic entity after the cloning, share the very same genotype. And although current cloning technologies require the fusion of skin cells with enucleated eggs, future technologies may render this fusion unnecessary: in such a case, the skin cell before the cloning, and the zygotic entity after the cloning, would share the very same parts. The short answer to this objection runs roughly as follows. The fact that x and y share the same genotype is not sufficient to render x and y numerically identical with one another. Different cells in my body share the same genotype but are not thereby numerically identical, and "identical" twins may share the same genotype without being numerically identical. The mere fact that x and y are *genetically* identical does not make x and y *numerically* identical.

Likewise, the fact that x and y share the same parts is not sufficient to render x and y numerically identical with one another. All the parts that constitute my body right now (x) will one day be dispersed here, there, and everywhere and will constitute a very scattered particular (y). But I am not numerically identical to that scattered particular. Conversely, there is right now a scattered particular (x) whose constituent parts will one day constitute my body (y), even though that scattered particular is not numerically identical to me. Nor would our judgments change if we factored out the "scattered" aspect of these examples: even if I truthfully believed that the parts constituting some (non-scattered) lump of stuff *right there* used to constitute me (or will one day constitute me), I would not be committed to the belief that I am numerically identical to this lump. The mere fact that x and y are *mereologically* identical (from the Greek *meros*, "part") does not make x and y *numerically* identical.

---

[35] Buckle (1990).

Finally, even if *x* and *y* share the same genotype *and* the same parts, this is still not sufficient to render *x* and *y* numerically identical with one another. Imagine *x* and *y* are adjacent, genetically identical cells in your body. Now imagine that the molecular turnover of *x* and *y* are such that the parts which constitute *x* right now will one day constitute *y*. (Whether or not this ever happens in the human body is beside the point, for we can imagine setting up an experiment to guarantee that it will happen.) This arrangement would not prompt us to say that *x* is numerically identical to *y*. More precisely, this arrangement would not prompt us to say that *x before the molecular turnover* is numerically identical to *y after the molecular turnover*. After all, *x* and *y* might both persist though their respective molecular turnovers; if so, we would want to identify *x before the molecular turnover* with *x after the molecular turnover*, just as we would want to identify *y before the molecular turnover* with *y after the molecular turnover*. I conclude that there is no good reason for viewing the individual skin cell *before a cloning intervention* as numerically identical to the individual human organism *after the cloning intervention*, and that this is so even if they share *both* the same genome *and* the same parts.

The second objection—corresponding to the second necessary condition for "development"—emerges in the following way. The strategy of classifying various instances of assistance by the *effects* of the assistance, rather than by the *kinds* of assistance, simply shifts the problem of *kinds* back a level. For this strategy presupposes that the entities undergoing those effects are different *kinds* of entities before the assistance is applied: the reason why the *effect* of somatic cell nuclear transfer is thought different than the *effect* of sustaining and implanting an in vitro embryo is that the somatic cell is thought to be a different *kind* of entity than the in vitro embryo to begin with. Likewise, the difference in kind between the somatic cell and the in vitro embryo is explained by yet another reliance on the notion of *kinds*: the biological entity before the somatic cell nuclear transfer is a *different kind* of entity than the biological entity after the somatic cell nuclear transfer; while the biological entity before the implantation of the in vitro embryo is the *same kind* of entity as the biological entity after implantation. The objection can be put modestly: it is not obvious why the individual skin cell that generates a zygote is a different *kind* of thing from the zygote. What is it that provides the *grounds* for classifying these entities as different (or the same) *kinds* of entity? Or the objection can be put more boldly: perhaps human skin cells *are* individual human organisms: they are, after all, *human* skin cells, not *sheep* skin cells.

The short answer to this objection begins by appealing to a more careful use of language. The skin cells are indeed *human*, but the cells are not *humans*. For words like "human", "cow", "lamb" (and more generally, for terms denoting what have traditionally been called *living substances*), there is an important distinction between a *count* sense and a *stuff* sense. Joshua Hoffman and Gary Rosenkrantz illustrate this distinction by pointing out an ambiguity in the phrase "Mary had a little lamb." When "lamb" is interpreted as a substance in the count sense, this phrase means something like "Mary owns a small lamb"; but when "lamb" is interpreted as a substance in the stuff sense, this phrase means something like "Mary ate a small amount of lamb".[36] The stuff sense is also sometimes called the *mass* or *quantity* sense, as explained by J.P. Moreland and Scott Rae:

> There are two different uses of *substance*: the count sense (in which a whole cow counts as one substance and of which we may ask, e.g., 'How many cows are in the field?') and the mass or quantity sense (in which the amount of a certain kind of material or stuff is in view and of which we may ask, e.g., 'How much cow did you eat, five or ten ounces?').[37]

This distinction between a count sense and a stuff (mass, quantity) sense has a clear relevance for marking the difference between skin cells and zygotes. Individual body cells in an adult, such as gametes or skin cells, are each human in the stuff sense but not in the count sense. But with the zygote, what we have is an individual body cell that is human *both* in the stuff sense *and* in the count sense. This is why skin cells are not human organisms, and hence why a skin cell is different in *kind* from an in vitro embryo. But here it will no doubt be asked: how can we make sense of this analysis of the zygote? How can it be human *both* in the stuff sense *and* in the count sense, even though every other bodily cell is human *only* in the stuff sense?

The justification for this analysis begins by reflecting on the fact that during every phase of our bodily existence, including the adult phase, we are composed of, or constituted by, human cells. These constituent cells, whether considered individually or whether considered as a conglomerate, are human in the stuff sense but not in the count sense. But surely, an adult human organism is human in the count sense. Reflection upon adult human organisms tells us something both straightforward and important: *an entity can be human in the count sense even if each and every one of the cells which constitute it are only human in the stuff sense*. But the human zygote is simply a limiting case of this rather straightforward idea. The zygote is composed of only one cell, so the phrase "each and every one of the cells which constitute it" picks out only one cell. The cell that constitutes the human zygote, *qua* cell, is only human in the stuff sense. But the human zygote itself, *qua* organism, is human in the count sense.

To recap: one answer to Q2 is that the in vitro embryo and the body cell are not to be distinguished by the *kinds* of assistance they need in order to eventually end up with a baby, but rather are to be distinguished by the *effects* of the assistance they need in order to eventually end up with a baby. The in vitro embryo has the potential to develop into or *become* a baby, while the somatic cell has at most the potential to generate or *produce* a baby. This answer receives further support when we reflect on the difference between being human in the count sense and being human in the stuff sense.

---

[36] Hoffman and Rosenkrantz (1997, pp. 74–75). They in turn credit Roderick Chisholm for first employing "this rather droll figure" in a 1973 Brown University metaphysics class: see p. 201, footnote 2.

[37] Moreland and Rae (2000, p. 350), footnote 31.

Do skin cells and zygotes have different "natures"? A different way of answering Q2, without appealing to the *effects* of the assistance, is to refer directly to the *nature* of the entities in question.[38] Perhaps this is why Charo, in suggesting the most likely answer to Q2, goes immediately to the concept of a thing's *nature*. She thinks the "real response" to Q2 is an argument "that the 'implicit nature" of an embryo is to divide and grow into a baby, whereas the implicit nature of a skin cell is to divide into more skin cells.[39] She then quickly provides four reasons for doubting the effectiveness of such an argument. I shall spend the remainder of this section examining these four reasons.

Objection 1: the "natural" endpoint of in vitro embryos. Charo's first reason for doubting that there is any difference between the "implicit nature" of skin cells and embryos is this:

> . . .it has been shown that the implicit nature of an in vitro embryo is merely to divide; absent extensive external assistance, a baby is not its natural endpoint, only a mass of nonviable cells.[40]

Her introductory phrase, "it has been shown that" refers to a passage found earlier in her essay. We have already looked at a segment of it above; the whole passage reads as follows:

> A fertilized egg or early embryo in a petri dish most certainly has an intrinsic tendency to continue growing and dividing. Without the provision of an artificial culture medium, however, it will never grow and divide more than about 1 week. If the provision of such a medium is considered a form of external assistance akin to that at issue in passive potentiality, then the fertilized egg is a potential week-old embryo, not a potential baby.[41]

When we combine these two passages, Charo's argument seems to be this: (first premise) we are justified in viewing an entity as endowed with the "implicit nature" of dividing and growing into a baby only if the "natural" result of this entity's division and growth *is* a human baby. But (second premise) the "natural" result of the division and growth of an in vitro embryo is *not* a human baby. Hence (conclusion) we are *not* justified in viewing an in vitro embryo as endowed with the "implicit nature" of dividing and growing into a baby.

This argument is valid, but unfortunately its second premise is false. The second premise is false because it assumes what is "natural" for a given thing is what *does* happen to that thing in the *actual* course of its history. But what is "natural" for a given thing cannot be determined merely by observing what the *actual fate* of that thing turns out to be. To see this, note that we are inclined to view our children as "naturally" developing into adults. But imagine that an environmental disaster removed a necessary factor in our child's development to adulthood—for example,

---

[38] In addition, this way of answering Q2 is also a way of answering the question about *kinds* without appealing to *stuff* and *count* senses of "human". For that matter, this way of answering Q2 is also a way of answering Q1 without appealing to "active" potential.

[39] Charo (2001, p. 86).

[40] Charo (2001, p. 86).

[41] Charo (2001, p. 86).

imagine that a huge meteor suddenly removed all the breathable oxygen from the atmosphere near the earth. The understanding of "natural" encapsulated by the second premise would commit us to now *denying* that our children "naturally" develop into adults. Since this commitment is mistaken, the second premise is mistaken as well. Furthermore, Charo has already put her finger on why the second premise would be rejected by an advocate of the AFP. Notice that her last quoted sentence is a conditional, and it is precisely the antecedent that a thoughtful person must deny: "*If* the provision of such a medium is considered a form of external assistance akin to that at issue in passive potentiality, *then* the fertilized egg is a potential week-old embryo, not a potential baby"[42]. But the provision of an artificial culture medium is *not* a form of external assistance akin to that at issue in passive potentiality. As we saw above, Charo explicitly recognizes this. Hence it is unclear why she thinks that appealing to this idea can undermine an "implicit nature" argument for distinguishing the zygote from the body cell.

Finally, it is important to note Charo's logical slide concerning a related trio of claims. She seems to think that an advocate of the "implicit nature" distinction is committed to the following two claims (my formulations, not hers):

(1) The implicit nature of an embryo is to divide and grow into a baby, given external factors *F* (where *F* includes "certain things to make that process [of growth] continue unimpeded, for example food and water").[43]

(2) The implicit nature of a skin cell is to divide (and grow) into more skin cells.

But then she attempts to undermine (1) by making the following claim:

(3) The implicit nature of an in vitro embryo is to divide into a mass of nonviable cells, "absent extensive external intervention".

Charo's suggestion here is that the nature of an in vitro embryo described in (3) is *more* like the nature of a skin cell described in (2) and *less* like the nature of a normal embryo described in (1). She wants to conclude from this suggestion that the distinction between (1) and (2) does not apply in the case of in vitro embryos.

But the initial suggestion is a mistake to begin with. Once we understand the phrase "absent extensive external intervention" in (3), it should be clear that the nature described in (3) *is not* like the nature described in (2) but *is* like the nature described in (1). Regardless of how "extensive" they are, the external interventions are merely trying to make the in vitro embryo's process of growth continue unimpeded. Hence "absent extensive external intervention" just means "absent external factors *F*". We can thus rephrase (3) as (3a):

(3a) The implicit nature of an in vitro embryo is to divide into a mass of nonviable cells, absent external factors *F*.

---

[42]My emphasis.

[43]Charo (2001, p. 86).

Once we see this, we also see that (3a) and (1) are simply two sides of the same biological coin. The implicit nature of *any* embryo (in vitro *or* in vivo), *absent* external factors *F*, is to divide into a mass of non-viable cells, while the implicit nature of *any* embryo (in vitro *or* in vivo), *given* external factors *F*, is to divide and grow into a baby.

Objection 2: the non-skin genes of skin cells are merely dormant. Charo's second reason for doubting that there is any difference between the "implicit nature" of skin cells and embryos is this:

> ...the 'implicit nature' of a skin cell is also to divide and grow. With the discovery that the genes coding for non-skin functions are merely dormant rather than dead comes the observation that the nature of skin cells and that of embryonic cells are not terribly different.[44]

Using the numbered claims introduced a moment ago, Charo's point is this: because the *entities* described in both (1) and (2) can both "divide and grow," and because the *entities* described in both (1) and (2) both have "dormant" rather than "dead" parts, the "implicit nature" described in (2) is "not terribly different" than the "implicit nature" described in (1).

We can begin our reply by noting that mere division and growth, by themselves, do not help to classify what it is that is undergoing the division and growth. Although many different kinds of animal can divide and grow, merely knowing that a given animal divides and grows does not help us know what *kind* of animal it is: a pig is not a platypus, even though both undergo division and growth. Likewise, although many different kinds of one-celled organisms *within* an animal can divide and grow, merely knowing that a given one-celled organism divides and grows does not help us know what *kind* of one-celled organism it is: a liver cell is not a cheek cell, even though both "divide and grow". Hence whether we are comparing whole animals with whole animals, or one part of an animal with other parts of that same animal, the basic idea is the same: mere division and growth are not enough to tell us what the nature of the thing in question is.

This basic idea continues to be relevant when we attempt to compare the parts of an animal with a whole animal: a platypus and a liver cell are not the same kind of thing, even though both divide and grow. More to the point, merely knowing that a given one-celled organism divides and grows does not help us know whether this one-celled organism is a *whole* organism in its own right or a mere *part* of another organism. I shall return to this idea below, but for now, suffice it to say that if this last idea is correct, then division and growth are not doing the real work in Charo's second objection.

What is doing the work in this second objection, I think, is the distinction between "dormant" and "dead". Charo claims that since "the genes coding for non-skin functions are merely dormant rather than dead," there is not much difference between an adult skin cell and a zygote. In assessing this claim, we should recall Charo's genetic application of the distinction between the reversible coma and the

[44] Charo (2001, p. 87).

permanent vegetative state. For I think that by using this distinction—and, by extension, the dormant/dead distinction under discussion—she has given herself a bit of an unearned rhetorical advantage. Describing the genes, which are *parts* of an organism, as if they were a *whole* organism in their own right, is what allows Charo to talk as if the difference between genes being permanently shut-off and genes being dormant, is really the same as the difference between a human in a permanent vegetative state and a human in a reversible coma.

Such talk, however, cannot be taken literally. Since *genuine* permanent vegetative states and *genuine* reversible comas are had by whole organisms, the sense in which genes can be said to be in these states is a very attenuated one. A whole organism, in the process of coming out of a reversible coma, meets the following three conditions: it

(1a) starts as a whole organism,
(2a) finishes as a whole organism, and
(3a) remains the self-same whole from the beginning of the process to the end of the process.

But the genes in a skin cell do not meet any of these three conditions. By their involvement in a process of somatic cell nuclear transfer, such genes

(1b) start as *a part of* a whole organism,
(2b) finish as *a part of* a whole organism, and
(3b) do *not* remain the self-same part from the beginning of the process to the end of the process.

Although (2b) and (3b) are both plausible and can be defended, (1b) is all we need to make a reply to Charo. For human rights belong to *whole* organisms and not to their parts. If I were to pluck out your eye, or steal some of your blood, I am not violating your *eye's* rights, or your *blood's* rights. I am violating *your* rights as a whole human organism. Hence when making the analogy between human genes and human patients in a reversible coma, we must be careful not to allow several of the associations we have with human patients to be smuggled in: the sense of hoping for recovery, the conviction that we have a present locus of value, the feeling of shock that we would have if the one in the coma were to be dismembered, and so on. These associations are entirely appropriate in the case of adult human organisms, but they are entirely inappropriate in the case of genes. In short, since the dormant/dead distinction is similar to the reversible coma/PVS distinction, while it may be true that both distinctions can be taken literally with *whole* organisms, neither distinction can be taken literally with *parts* of organisms (like genes).

There is, however, another way of understanding Charo's comparison. Perhaps the relevant biological unit she is trying to compare with the zygote is not the *genes* of the somatic cell, but the somatic cell *itself*, which *has* those genes. After all, those who wish to protect human embryos from destructive experiments are not merely trying to protect the *genes* of those embryos, which are merely parts; rather, the biological unit in focus is the *whole* biological unit which *possesses* the genes—in this case, the zygote itself. Likewise, perhaps what Charo wishes to attend to, in the case of somatic cell nuclear transfer, is the *whole* biological unit which *possesses* the genes—the somatic cell itself.

The basic idea, then, would be something like this: just as there is a sort of genetically-based dormancy in the early embryo, so too there is a sort of genetically-based dormancy in the somatic cell. Hence if we are willing to grant a certain level of protection to the early embryo on account of *its* genetically-based dormancy, we seem to be committed to giving the same level of protection to the somatic cell on account of *its* genetically-based dormancy. In both cases, a one-celled being is under discussion. In both cases, the one-celled being can "divide and grow". And in both cases, the one-celled being contains genes, some of which are functioning and some of which are dormant. Even if we grant that the PVS/reversible coma distinction is only loosely applicable to the *genes*, an important question still remains: once we admit that a one-celled being like a human zygote can have moral status on account of its genetically-based dormancy, how are we to block the implication that *all* one-celled beings with genetically-based dormancy can have moral status?

The answer to this question will revolve around the different sorts of dormancy possessed by the genes in the somatic cell and the genes in the zygote. These different sorts of dormancy are exhibited by the different senses in which the cells in question "divide and grow". Strictly speaking, the skin cell divides and is no more, being replaced, as it were, by its two daughter cells. Hence the skin cell does not really "grow" even though the *mass* or *quantity* of skin cells does grow. Now in one sense, the same sort of fate is had by the zygote. After all, the zygote does divide, and in the process of this division, one cell gets replaced by two cells. But in another sense, the zygote remains despite its division: its "replacement" amounts to no more than one form of an organism's body replacing another form of that *same* organism's body. In this sort of "replacement", the organism constituted by the one cell *endures* through the change in cell number, and this change simply *endows*, as it were, the already-existing organism with two cells instead of one. From the vantage point of the cell, the change is a *replacement*, but from the vantage point of the organism, the change is merely a *re-constitution*.

Objection 3: the transformation from embryo to skin cell is gradual. Charo's third reason for doubting that there is any difference between the "implicit nature" of skin cells and embryos is this:

> This observation [that the nature of skin cells and that of embryonic cells are not terribly different] is made more acute by noting that the transformation of cells from embryonic to fetal to adult is one marked by a gradual decline in the functioning of their nonspecialized genes; the distinction between these stages of cell life is not clean but blurred.[45]

I shall not have a great deal to say in response to this reason. Charo does not attempt to spell out how the *graduality* of the decline in nonspecialized gene functioning adds anything to her second objection. Since the main claim of the second

[45]Charo (2001, p. 87).

objection was that the nature in (2) was not so different than the nature in (1), perhaps the basic idea of this third objection is as follows: if x and y stand at opposite ends of an ordered succession of things, and if each thing in this succession is only slightly different than the things next to it, this makes it more difficult to claim that x and y have different natures.

If this is the basic idea behind the objection, then it can be answered by presenting a point Michael Tooley once made in his criticism of the graduality-based argument for a conservative position on abortion. Tooley asks us to consider:

> a series of objects, starting with a very heavy one, followed by one that is a hundredth of a gram lighter, followed by another that is another hundredth of a gram lighter, and so on down to an object that is very light indeed. It is impossible to find two successive objects, one of which is heavy, and the other not.[46]

Since this arrangement would not lead us to conclude that *all* of the objects in the series are heavy, this shows that:

> the absence of significant differences between successive members of some series, or between successive stages in some process, provides no reason at all for concluding that there are no significant differences between non-successive stages or members.[47]

The relevance of this point to the present issue should be clear. Consider a series of cells, starting with a cell whose nonspecialized genes are functioning at peak capacity, followed by a cell whose nonspecialized genes are functioning slightly below peak capacity, and so on, down to a cell whose nonspecialized genes are hardly functioning at all. The absence of significant differences between successive cells in this series, in terms of the functioning of their nonspecialized genes, provides no reason at all for concluding that there are no significant differences between non-successive cells in this regard. Hence the claim that the skin cell and the zygote have different natures is not threatened by the gradual decline Charo notes.

Objection 4: the possibility of cytoplasmic reprogramming. Charo's fourth reason for doubting that there is any difference between the "implicit nature" of skin cells and embryos is this:

> ...when current research on somatic cell nuclear transfer cloning unlocks the secrets to the role of the egg's cytoplasm in regulating the expression of genes that code for development of an entire organism, the next step in cloning will be to eliminate the need for fusion with an enucleated egg; all the material needed to regulate gene expression is present in the cytoplasm of the skin cell as well, and turning that expression on by manipulation of the skin cell alone will be the final step in eliminating all pertinent difference between embryonic cells and adult cells.[48]

Charo thinks these interventions into the life of the cell, should they become a technological possibility, will deplete even further the distinction between a human zygote and a human somatic cell. There are (at least) two ways of envisaging this,

---

[46] Tooley (1983, p. 170).

[47] Tooley (1983, p. 170).

[48] Charo (2001, p. 87).

and her passage, as stated, does not clearly commit her to just one of them. First, perhaps we could *inject* somatic cells with the substances found in egg cytoplasm that trigger growth and differentiation, so that these *injected* materials trigger growth and differentiation in the somatic cell. Second, perhaps we could manipulate the *existing* cytoplasmic materials, without injecting any new substances, so that these *existing* materials trigger growth and differentiation. In order to allow Charo's point here to be as strong as possible, let us assume that the second of these scenarios has become a technological possibility. If even this second scenario does not present a serious obstacle to viewing the zygote and the skin cell as endowed with different "implicit natures", then neither does the first scenario.

My reply is threefold. First, I have already explained above how *mereological* identity is not sufficient for *numerical* identity: x and y can have all the same parts and still be numerically distinct. Now the same consideration applies to the *natures* of x and y: x and y can have all the same parts and still have different natures. Recall the above discussion of the non-scattered lump of stuff that used to constitute my body: I do not have the same nature as that lump, even though all the parts that constitute the lump once constituted me.

Second, the last clause in Charo's passage inadvertently lends support to the view she is critiquing: "turning that expression on by manipulation of the skin cell alone will be the final step in eliminating all pertinent difference between embryonic cells and adult cells". She wants to suggest that the possibility of such manipulation implies that the adult cell was relevantly similar to the embryonic cell even *before* the manipulation took place. But this amounts to simply *ignoring* the "pertinent difference" that existed before the manipulation took place. "Eliminating" this pertinent difference by manipulating the skin cell is precisely what is required in order to *transform* the skin cell into an embryonic cell.

Third, the considerations Charo introduces here are relevantly similar to the considerations involved in certain *other* cases of embryogenesis, whereby one part of a biological organism, without undergoing fusion, nevertheless gives rise to a new, distinct organism in its own right. Since these *other* cases of embryogenesis do not require us to view the part in question as endowed with the same nature as the embryo that emerges from that part, neither does the case of cytoplasmic reprogramming. Examples of such embryogenesis include the *fission* of the early embryo, whether occurring naturally (in which case it is called *monozygotic twinning*) or artificially (in which case it is called *blastomere separation* or *embryo splitting*) and *parthenogenesis*, a form of reproduction in which the ovum gives rise to a new biological individual without being fertilized.[49]

Two decades ago, Warren Quinn wrote that parthenogenesis could be viewed as "a transformation rather than a development of the ovum," whereby the parthenogenic agent is seen as "having the power to change the ovum's essential nature, to make of it a new organism with a quite different teleology". He noted

[49] Charo discusses embryo splitting Charo (2001, pp. 87–88).

that cloning could be viewed in the same way.[50] More recently, David Oderberg discusses cloning in the context of his analysis of parthenogenesis, twinning, and the totipotency of individual cells making up an early (multicellular) embryo. Oderberg's overall goal is to assimilate all kinds of embryogenesis to "a natural class of events that involve the changing of an entity which is part of the cause of the emergence of a new human being into an entity which *constitutes* that new human being."[51] Of particular relevance to the present point is his treatment of the totipotent cells making up the early embryo. He argues that the *potential* of each cell in an early embryo to eventually give rise to a distinct human individual (via embryo splitting) is not the same as each cell *actually being* a distinct human individual "while it subserves the embryo of which it is a part."[52] Applying this idea to the present point, the idea is that the *potential* of each cell in an adult to give rise to a distinct human individual (via cytoplasmic reprogramming) is not the same as each cell's *actually being* a distinct human individual while it subserves the adult of which it is a part.

Hence regardless of what sort of cell we begin with, and regardless of what event *triggers* the embryogenesis, the main point is the same: "cells can be altered intrinsically, by events which scientists are becoming ever more competent to identify, from undetached constituent parts of the organism, to detached embryos in their own right."[53] The possibility of cytoplasmic reprogramming does not threaten this basic insight, and hence does not threaten the idea that the implicit nature of skin cells and embryos are different.

In summary, Charo argues that in an age of cloning such as ours, there is not a morally relevant difference between the sort of potential possessed by a skin cell and the sort of potential possessed by an embryo. But her suggestion that both an in vitro embryo and a somatic cell possess the relevant "active" potential neglects the fact that the *effect* of the assistance is very different between the two cases. Likewise, the reasons she gives for doubting any differences between the *nature* of skin cells and zygotes are ultimately unconvincing. Even in an age of cloning, the claim that some cells are "sacred" because of their potential does not entail the claim that every cell is sacred.

3 Potential Presidents and Potential Persons

As was mentioned at the outset of this chapter, the AFP is often formulated using the language of rights. When formulated this way, the AFP is often opposed by the

[50] Quinn (Winter 1984, p. 28).

[51] Oderberg (1997, p. 297). The meaning of "human being" for Oderberg is the same as the meaning of "human organism" in my discussion above.

[52] Oderberg (1997, p. 280).

[53] Oderberg (1997, p. 292).

example of the potential president: just as the mere fact that an individual is a potential president does not give that individual the right to command the military, so too the mere fact that an individual is a potential person does not give that individual the right to life. In this section, I will argue that the example of the "potential president" alerts us to a possible mistake to avoid in formulating the AFP. But I will also argue that the formulation of the AFP represented by Chapters 1 through 3 of this book does not make that mistake.

The following claim is often invoked in discussions about the morality of abortion, infanticide, embryonic stem cell research, and a number of other issues that touch upon the beginning stages of human life:

> POTENTIAL RELEVANT TO MORAL STATUS: an entity's potential is relevant to its moral status

One of the more common ways of fleshing out this claim relies upon the concept of a potential person and the concept of a right to life, as follows:

> POTENTIAL PERSONHOOD GENERATES A RIGHT TO LIFE: the fact that an entity is a potential person is sufficient for it to possess a right to life

This latter claim, in turn, is commonly opposed by the example of the potential president, which received its initial formulation by the Australian philosopher Stanley Benn:

> [My argument] is not the argument that infants are *potential* persons, and have rights as such. For if *A* has rights only because he satisfies some condition *P*, it doesn't follow that *B* has the same rights now because he *could* have property *P* at some time in the future. It only follows that he *will* have rights *when* he has *P*. He is a potential bearer of rights, as he is a potential bearer of *P*. A potential president of the United States is not on that account Commander-in-Chief.[54]

Benn's example of the potential president has received a fair amount of philosophical attention over the last thirty years, and variations on the example abound—for example, Peter Singer used a distinctively British version of the example in his *Practical Ethics*, noting that "Prince Charles is a potential King of England, but he does not now have the rights of a king."[55]

It is important to have a clearer understanding of the operative concept of a person that appears both in POTENTIAL PERSONHOOD GENERATES A RIGHT TO LIFE and in the example of the potential president, since this operative concept of a person is importantly different than the concept of a person I have been using throughout this book. Indeed, one of the reasons that I have chosen to formulate the thesis of this book in a way different than POTENTIAL PERSONHOOD GENERATES A RIGHT TO LIFE is that I believe POTENTIAL PERSONHOOD GENERATES A RIGHT TO LIFE relies upon what I take to be an unusual and inadequate concept of a person. The concept of a person in POTENTIAL PERSONHOOD GENERATES A RIGHT TO LIFE, even when interpreted charitably, is still somewhat technical, narrow, and

---

[54] Benn (1984, p. 143).

[55] Singer (1993, p. 153).

apt to mislead. Still, it is worth investigating whether the potential president example really does count against POTENTIAL PERSONHOOD GENERATES A RIGHT TO LIFE, as stated, before moving on to investigate whether the potential president example might count against the version of the AFP represented by Chapters 1 through 3 of this book. And the first step in this investigation is getting clear on what might be meant by "person" and "potential person" in POTENTIAL PERSONHOOD GENERATES A RIGHT TO LIFE.

Recall from Chapter 1 how one of the most important distinctions in the way "person" is used is what Joel Feinberg calls the distinction between normative (or moral) personhood on the one hand and descriptive (or commonsense) personhood on the other:

> To be a person in the *normative* sense is to have rights, or rights and duties, or at least to be the sort of being who could have rights and duties without conceptual absurdity...when we attribute personhood in a purely *normative* way to any kind of being, we are attributing such moral qualities as rights or duties, but not (necessarily) any observable characteristics of any kind—for example, having flesh or blood, or belonging to a particular species.[56]

> There are certain characteristics that are fixed by a rather firm convention of our language such that the general term for any being who possesses them is "person."...I shall call the idea defined by these characteristics "the *commonsense* concept of personhood." When we use the word 'person' in this wholly *descriptive* way we are not attributing rights, duties, eligibility for rights and duties, or any other normative characteristics to the being so described. At most we are attributing characteristics that may be a *ground* for ascribing rights and duties.[57]

POTENTIAL PERSONHOOD GENERATES A RIGHT TO LIFE uses "person" in a descriptive sense, not a normative sense. The claim being made by POTENTIAL PERSONHOOD GENERATES A RIGHT TO LIFE is *not* that the fact that an entity is a potential *normative* person is sufficient for it to possess a right to life; rather, the claim is that the fact that an entity is a potential *descriptive* person is sufficient for it to possess a right to life. POTENTIAL PERSONHOOD GENERATES A RIGHT TO LIFE, as stated, leaves open the question of precisely which characteristics—reason, and/or consciousness, etc.—are constitutive of being a "person" in a descriptive sense. Indeed, one might think that the characteristics constitutive of being a "person" in a descriptive sense are certain *capacities*—understood, perhaps, as *immediate* capacities, or then again, perhaps as *first-order* capacities. However, as it is usually understood, POTENTIAL PERSONHOOD GENERATES A RIGHT TO LIFE does not usually make room for the idea that the characteristics constitutive of being a "person" in a descriptive sense are certain *higher-order capacities of some level or other.* Still, the present point is that POTENTIAL PERSONHOOD GENERATES A RIGHT TO LIFE uses "person" in a descriptive, and not a normative, sense.

The example of the potential president also uses "person" in a descriptive sense, not a normative sense. As Benn originally put it,

---

[56]Feinberg (1980, p. 186), emphasis mine.

[57]Feinberg (1980, p. 187), first two emphases mine.

> I characterize a person. . .as someone aware of himself, not just as process or happening, but as *agent*, as making decisions that make a difference to the way the world goes, as having projects that constitute certain existing or possible states as "important" and "unimportant," as capable, therefore, of assessing his own performances as successful or unsuccessful.[58]

One need not be committed to Benn's proposal of the characteristics constitutive of being a "person" in order to rely upon his example of the potential president. However, one certainly does need to keep to *some* descriptive use of "person"—and to stay away from *all* normative uses of "person"—in order to rely upon Benn's example. Otherwise, the entire structure of the example, and its relevance to POTENTIAL PERSONHOOD GENERATES A RIGHT TO LIFE, collapses.

One of the standard responses to Benn's example of the potential president is to claim that, although he is correct both in his description of the example and in the general principle he extracts from the example, he is mistaken in thinking that either the general principle or the example are relevant to POTENTIAL PERSONHOOD GENERATES A RIGHT TO LIFE. According to this standard response, Benn has simply misunderstood the defender of POTENTIAL PERSONHOOD GENERATES A RIGHT TO LIFE. I believe this standard response is correct, and showing why it is correct also shows why the argument of Chapters 1 through 3 of this book is not vulnerable to Benn's objection.

First, consider why the standard response is correct. Benn imagines the defender of POTENTIAL PERSONHOOD GENERATES A RIGHT TO LIFE arguing as follows (for the sake of clarity, I will drop Benn's use of the labels "A" and "B", and will insert the rider "at [time] t"):

(1) X has the right to life at t only because X is a person at t.

Therefore,

(2) X has the right to life at t because X could be a person at some time after t.

If the defender of POTENTIAL PERSONHOOD GENERATES A RIGHT TO LIFE really were arguing from (1) to (2), then the example of the potential president would be an appropriate criticism. As Joel Feinberg put it,

> It is a logical error. . .to deduce *actual* rights from merely *potential* (but not yet actual) qualification for those rights. What follows from potential qualification. . .is potential, not actual, rights; what entails actual rights is actual, not potential, qualification. As the Australian philosopher Stanley Benn puts it, "A potential president of the United States is not on that account Commander-in Chief [of the U.S. Army and Navy]." This simple point can be called "the logical point about potentiality."[59]

However, a standard response to Benn's example is that Benn has misunderstood the defender of POTENTIAL PERSONHOOD GENERATES A RIGHT TO LIFE. After all, this standard response goes, the defender of POTENTIAL PERSONHOOD

[58] Benn (1984, p. 141).

[59] Feinberg (1980, p. 194).

GENERATES A RIGHT TO LIFE would not accept (1) in the first place. Still less would she attempt to argue for (2) on the basis of (1). As Michael J. Wreen notes,

> Proponents of [the view that all potential persons have a right to life] would agree that mere potential possession of the qualifications for a right is not sufficient for actual possession of that right. But the question, they would add, is what the actual qualifications for possession of a right to life are. They hold that potential personhood is that—or, better, one such—actual qualification, or sufficient condition, and so quite correctly claim that the "logical point about potentiality" that Feinberg mentions counts nought against them.[60]

I believe this standard response is correct. And this standard response is relevant to the main argument of this book, even though the main argument does not make use of the concept of "potential persons." The argument of this book is not: because an individual's possession of a certain set of immediate capacities generates serious moral status for that individual, *it follows that* an individual's possession of the capacity to have these immediate capacities generates such moral status. Rather, the argument of this book is: since there are cases where an individual's possession of certain higher-order capacities *by itself* generates serious moral status for that individual, it follows that possession of these higher-order capacities is always sufficient to generate such moral status. Hence, the argument of this book does not make the logical mistake that Feinberg borrows from Benn's example of the potential president.

However, I recognize that there is another way of employing Benn's example that might appear to cause problems for the argument of this book. In particular, the example of the potential president does more than merely illustrate the fallacy of reasoning from (1) to (2), or "the logical point about potentiality". In addition, the example of the potential president subtly advances a certain way of thinking about persons in the first place, and it is this way of thinking about persons that might appear to cause problems for the argument of this book.

To see why this is so, notice that Benn's example of the potential president relies upon two assumptions about presidents and two parallel assumptions about persons. While the assumptions about presidents are unobjectionable, the assumptions about persons are much more objectionable. These "loaded" assumptions, as I shall provocatively label them, do much work for Benn (and countless others who have never heard of Benn) in "stacking the deck" against the sort of position I argue for in this book.

Benn's first unobjectionable assumption is that the term "president" is what logicians call a *phase sortal* rather than a *substance sortal*:

> BENN'S FIRST UNOBJECTIONABLE ASSUMPTION: The term "president" is a phase sortal.

This assumption is rarely made explicit, but it is important to do so. In Chapter 1 I explained how *sortals* are count-nouns that serve as classificatory concepts for describing the world, and they can be divided up into *substance* sortals and *phase*

---

[60] Wreen (1986, p. 138).

sortals. The explanation of David Wiggins, cited earlier, deserves to be repeated again. He notes that the difference between substance sortals and phase sortals is

> between sortal concepts which present-tensedly apply to an individual *x* at every moment throughout *x*'s existence, e.g. *human being*, and those which do not, e.g. *boy*, or *cabinet minister.*[61]

Part of the logic of phase sortals is illustrated by Wiggins' example of the boy: there is nothing incoherent about an individual falling under a phase sortal *P* during one period of time, and yet failing to fall under *P* during a later period of time. Another part of the logic of phase sortals is illustrated by Wiggins' example of the prime minister: there is nothing incoherent about an individual failing to fall under a phase sortal *P* during one period of time, and then coming to fall under *P* during a later period of time. This is why it makes perfect sense to treat the term "president" as a phase sortal. There is nothing incoherent about an individual failing to be a president during one period of time, and then being a president during a later period of time.

Benn's second unobjectionable assumption is that having the right to command the military at a given time requires being an actual president at that time:

> BENN'S SECOND UNOBJECTIONABLE ASSUMPTION: Having the right to command the military at a given time requires being an actual president at that time.

When these two unobjectionable assumptions are combined, the result is that whenever an individual does not fall under the phase sortal "president", that individual does not have the right to command the military. And since, when an individual is merely a potential president, that individual does not fall under the phase sortal "president", it follows that, when an individual is merely a potential president, that individual does not have the right to command the military.

There are two "loaded" assumptions that must be accepted in order to make the example of the potential president relevant to discussions about potential persons. These two "loaded" assumptions are almost exactly the same as the two unobjectionable assumptions just considered, with the only differences being the substitution of the word "person" for the word "president" and the substitution of the phrase "right to life" for the phrase "right to command the military" in the relevant places:

> BENN'S FIRST LOADED ASSUMPTION: The term "person" is a phase sortal.
> BENN'S SECOND LOADED ASSUMPTION: Having the right to life at a given time requires being an actual person at that time.

When these two loaded assumptions are combined, the result is that whenever an individual does not fall under the phase sortal "person", that individual does not have the right to life. And since, when an individual is merely a potential person, that individual does not fall under the phase sortal "person", it follows that, when an individual is merely a potential person, that individual does not have the right

[61] Wiggins (1967, p. 7, 1980, 2001).

to life. If Benn's two loaded assumptions are true, then POTENTIAL PERSONHOOD GENERATES A RIGHT TO LIFE must be rejected as false.

But are Benn's two loaded assumptions true?

Some philosophers have attacked BENN'S FIRST LOADED ASSUMPTION by arguing that the term "person" is not a phase sortal at all, but is rather a substance sortal. This attack often takes the form of a complaint that there is something fundamentally incoherent in talk about a "potential person", or that to speak of a "potential person" is to make some kind of category mistake.

I am rather sympathetic with this line of attack. However, there are two reasons why, for the sake of the present argument, I am willing to grant BENN'S FIRST LOADED ASSUMPTION. The first reason is simply a matter of usage: it seems to me that the term "person" is sometimes used as a phase sortal, sometimes used as a substance sortal, and sometimes used with no clear commitment either way. Among philosophers and non-philosophers alike, these different uses of "person" often pass one another unnoticed, like ships in the night. While this practice of multiple usage may be regrettable, it is no good to deny that it goes on.

The second reason for granting BENN'S FIRST LOADED ASSUMPTION is simply a matter of consistency: it would be self-defeating for a defender of POTENTIAL PERSONHOOD GENERATES A RIGHT TO LIFE to complain that Benn uses "person" as a phase sortal. After all, POTENTIAL PERSONHOOD GENERATES A RIGHT TO LIFE uses "person" as a phase sortal too. Those philosophers who insist that "person" is a substance sortal can profitably argue, I believe, that this is an excellent reason to reject POTENTIAL PERSONHOOD GENERATES A RIGHT TO LIFE in favor of a more philosophically sophisticated formulation of POTENTIAL RELEVANT TO MORAL STATUS (the claim, recall, that "an entity's potential is relevant to its moral status"). I think such philosophers would be quite correct about this. Indeed, my conviction that they would be correct about this is one of the reasons I wrote this book without relying on the idea of a "potential person".

Still, I'm willing for the sake of discussion to pretend that "person" is a phase sortal, like "president". The question I intend to investigate is this: once BENN'S FIRST LOADED ASSUMPTION is granted, how plausible is BENN'S SECOND LOADED ASSUMPTION?

I believe this question can be investigated by focusing on a seldom-noticed feature of the American presidency: an individual can be a president during one period of time, cease being a president during a second period of time, and then become a president, again, during a third period of time.

One source of this feature of the American presidency is the provision in the American constitution that allows for presidents to run for re-election to a second term of office. Since nearly all first-term presidents have attempted to get re-elected for a second term, and since these re-election attempts are very much in the public view, it is somewhat surprising that philosophers have overlooked this feature of the presidency.

However, there may be an explanation available for why philosophers have overlooked this feature of the presidency: it is very unusual for a first-term president to even make the attempt at re-election for a second term, *after having taken a break from the presidency*, and it is even more unusual for such an attempt at re-election to succeed. This is not surprising, since the reasons why a first-term president would take a break from the presidency to begin with—illness, age, unpopularity, scandal, defeat, and death—are often excellent reasons for *not* running for the office in a future election.

Still, there is nothing incoherent about this feature of the American presidency. This feature of the American presidency illustrates how there is nothing incoherent about an individual falling under a phase sortal *P* during one period of time, failing to fall under *P* during a second period of time, and then falling under *P* again during a third period of time.

Imagine the following scenario. Al is president from 3000 to 3004, ceases being president in 3005, and then gets re-elected to be president from 3009 to 3012. Put a bit differently, Al is an actual president from 3000 to 3004, a potential president from 3005 to 3008, and an actual president from 3009 to 3012. The important point for present purposes is that from 3005 to 3008, Al is not an actual president but a potential president.

Now enter BENN'S SECOND UNOBJECTIONABLE ASSUMPTION. Since having the right to command the military at a given time requires being an actual president at that time, and from 3005 to 3008 Al is not an actual president, it follows that from 3005 to 3008 Al does not have the right to command the military.

The result illustrated by Al's case may be called the innocuous implication:

> THE INNOCUOUS IMPLICATION: If an individual temporarily fails to fall under the phase sortal "president" at a given time, that individual loses the right to command the military at that time.

In order to see how this example of re-electing the potential president has a parallel in the case of persons, we need only combine BENN'S FIRST LOADED ASSUMPTION with the truth, mentioned just a moment ago, that there is nothing incoherent about an individual falling under a phase sortal *P* during one period of time, failing to fall under *P* during a second period of time, and then falling under *P* again during a third period of time.

Imagine the following scenario. Bill possesses the properties that constitute being a person from 3000 to 3004, loses those properties in 3005, and then regains those properties from 3009 to 3012. Put a bit differently, Bill is an actual person from 3000 to 3004, a potential person from 3005 to 3008, and an actual person from 3009 to 3012. The important point for present purposes is that from 3005 to 3008, Bill was not an actual person but a potential person.

Now enter BENN'S SECOND LOADED ASSUMPTION. Since having the right to life at a given time requires being an actual person at that time, and since from 3005 to 3008 Bill was not an actual person, it follows that from 3005 to 3008 Bill does not have the right to life.

The result illustrated by Bill's case may be called the embarrassing entailment:

> THE EMBARRASSING ENTAILMENT: If an individual temporarily fails to fall under the phase sortal "person" at a given time, that individual loses the right to life at that time.

To see why this entailment is embarrassing, take any given definition of "person" in a descriptive sense: for example, a definition which claims that X is a person just in case X has reason. Now imagine that Bill has reason from 3000 to 3004, loses his reason from 3005 to 3008 due to a brain injury, and gets his reason back again in 3009. Benn's two loaded assumptions, when combined, entail that Bill does not have the right to life from 3005 to 3008.

THE EMBARRASSING ENTAILMENT can be reached even if the definition of person is changed to emphasize something other than reason: for example, a definition which claims that X is a person just in case X has consciousness. Imagine that Bill has consciousness from 3000 to 3004, loses his consciousness from 3005 to 3008 due to a brain injury (he exists, let's say, in a temporary coma), and gets his consciousness back again in 3009. Once again, Benn's two loaded assumptions, when combined, entail that Bill does not have the right to life from 3005 to 3008.

THE EMBARRASSING ENTAILMENT becomes all the more embarrassing when one realizes that the lapse in personhood is just as detrimental to Bill's right to life whether it lasts for 4 years, 4 days, or 4 minutes. Imagine Bill goes in for minor surgery, and is given a general anesthetic that makes him completely unconscious during the surgery. Benn's two loaded assumptions entail that Bill does not have the right to life during the time of this minor surgery.

There are at least three approaches for avoiding THE EMBARRASSING ENTAILMENT. The first is to claim that, even though Bill is not *actually* a person from 3005 to 3008, Bill still retains his right to life from 3005 to 3008, since Bill is a *potential* person during this time. According to this approach, an entity has the right to life at a given time as long as it is *either* an actual person at that time *or* a potential person at that time.

Unfortunately, this approach abandons BENN'S SECOND LOADED ASSUMPTION ("Having the right to life at a given time requires being an actual person at that time"). Indeed, this approach is tantamount to endorsing POTENTIAL PERSONHOOD GENERATES A RIGHT TO LIFE.

A second approach for avoiding THE EMBARRASSING ENTAILMENT is to claim that, even though Bill is not actually a person from 3005 to 3008, Bill still retains his right to life from 3005 to 3008 because he *already was* a person from 3000 to 3004. According to this approach, an entity has the right to life at a given time as long as it is *either* an actual person at that time *or* was an actual person at some previous time.

This second approach, unlike the first, is *not* tantamount to endorsing POTENTIAL PERSONHOOD GENERATES A RIGHT TO LIFE. Unfortunately, this second approach, just like the first, must abandon BENN'S SECOND LOADED ASSUMPTION. The example of re-electing the potential president shows why. Al did not have the right to command the military from 3005 to 3008, even though Al *already was* a president from 3000 to 3004. If Benn's two loaded assumptions are correct, there is no reason to think the situation is any different for Bill and the right to life than it is for Al and the right to command the military. Once BENN'S SECOND LOADED ASSUMPTION is abandoned, the alleged parallel between presidents and persons collapses, and the example of the potential president does not give us any reason for rejecting POTENTIAL PERSONHOOD GENERATES A RIGHT TO LIFE.

A third approach for avoiding THE EMBARRASSING ENTAILMENT is to invoke the moral and metaphysical claims I have been defending in this book: to claim that Bill does not cease being a person from 3005 to 3008, since the properties constitutive of being a person include both what I have called immediate capacities and what I have called higher-order capacities. For example, if we focus on a definition of a person which emphasizes reason, then, according to this approach, something is a person at a given time as long as it has *either* the immediate capacity to reason at that time *or* a higher-order capacity to reason at that time. The higher-order capacity to reason just is the ability to acquire the immediate capacity to reason. According to this approach, even though Bill does not have the immediate capacity to reason from 3005 to 3008, he still has the higher-order capacity to reason from 3005 to 3008. Therefore, Bill is still a person from 3005 to 3008.

This third approach, unlike the first one, does *not* require abandoning BENN'S SECOND LOADED ASSUMPTION. However, this third approach has an interesting result: without explicitly endorsing POTENTIAL PERSONHOOD GENERATES A RIGHT TO LIFE, this third approach ends up generating a right to life for precisely the same entities as POTENTIAL PERSONHOOD GENERATES A RIGHT TO LIFE. This is because the talk about higher-order capacities is simply another way of talking about potential. Saying that Bill has, from 3005 to 3008, the higher-order capacity to reason, is the same as saying that Bill has, from 3005 to 3008, the potential to reason.

Once it is recognized that this third approach depends upon potentiality in this way, it becomes clear that it is a mere notational variant on the first approach. The first approach combined two claims (focusing again on a reason-based definition of person):

1. X has the right to life only if X is a person *or a potential person.*
2. X is a person just in case X has reason.

The third approach takes these same two claims, and simply relocates the concept of potential (which it calls a higher-order capacity) from the first claim to the second claim:

1. X has the right to life only if X is a person
2. X is a person just in case X has reason *or the potential to have reason*

This third approach, then, is merely a slightly different way of fleshing out the claim labeled POTENTIAL RELEVANT TO MORAL STATUS above:

> POTENTIAL RELEVANT TO MORAL STATUS: An entity's potential is relevant to its moral status.

While POTENTIAL PERSONHOOD GENERATES A GIGHT TO LIFE is one way of fleshing out this claim, this third approach for avoiding THE EMBARRASSING ENTAILMENT suggests another:

PERSONHOOD* GENERATES A RIGHT TO LIFE: The fact that an entity is a person* (an entity with reason *or the potential to have reason*) is sufficient for it to possess a right to life.

POTENTIAL PERSONHOOD GENERATES A RIGHT TO LIFE and PERSONHOOD* GENERATES A RIGHT TO LIFE are extensionally equivalent: any entity that has a right to life according to one principle will also have a right to life according to the other principle.

I have just examined three ways of avoiding PERSONHOOD* GENERATES A RIGHT TO LIFE while retaining a commitment to BENN'S FIRST LOADED ASSUMPTION. Each of these ways involves either abandoning BENN'S SECOND LOADED ASSUMPTION or else involve packing potentiality into the concept of a person to begin with. If BENN'S SECOND LOADED ASSUMPTION is abandoned, the alleged parallel between persons and presidents collapses, and the example of the potential president does not give us any reason for rejecting POTENTIAL PERSONHOOD GENERATES A RIGHT TO LIFE. If potentiality is packed into the concept of a person to begin with, then there will be no difference in the range of entities that have a right to life according to POTENTIAL PERSONHOOD GENERATES A RIGHT TO LIFE and the range of entities that have a right to life according to PERSONHOOD* GENERATES A RIGHT TO LIFE.

There are three general lessons to be learned from this discussion of Benn's example of the potential president, and these lessons apply whenever an account of moral status uses the term "person". First, it is important to be clear on whether "person" is being used in a normative sense or in a descriptive sense. Second, it is important to be clear on whether "person" is to be taken as a phase sortal or a substance sortal. Third, whenever an account of moral status uses the term "person" in the *descriptive* sense as a *phase* sortal, it is important to be clear on whether X being a *potential* descriptive person is sufficient for X being a *normative* person. For example, if an account claims that X is a person just in case X has reason, does the account also claim that X being a potential person is sufficient for X having the right to life? If not, then one should expect the account of moral status in question to be vulnerable to the problem of re-electing the potential president.

Fortunately, the argument of this book takes these three lessons very seriously. First, I have used the term "person" in a descriptive sense rather than a normative sense: persons are things like you, and thinking about what might happen to you is an excellent way of thinking about what might happen to a person. Second, I have also used "person" as a substance sortal: you can stop being a number of things and still exist—rational, conscious, president, and so on—but you can never stop being a person and still exist, since you can never stop being you and still exist. Finally, I can happily avoid the fancy footwork of potential personhood while still noting how the metaphysical and moral framework I defend fits seamlessly with those who do embark on the path of potential personhood. The third lesson above can be translated to my account as follows. My account uses the phrase "normal adult human" in a *descriptive* sense as a *phase* sortal. X being a potential normal adult human is sufficient for X having serious moral status. Therefore, my account is not vulnerable to the problem of re-electing the potential president.

I noted at the start of this chapter that if the arguments of Chapters 2 and 3 were sound, then human organisms at the very earliest stages of their existence have the typical human capacities that are sufficient to generate serious moral status. These arguments thus constitute, among other things, a contemporary version of the Argument From Potential. The Argument From Potential, however, has frequently been charged with implying philosophically absurd conclusions and with depending on serious mistakes in moral reasoning. But this chapter has shown these charges to be mistaken, at least when hurled at the arguments of Chapters 2 and 3. Those arguments do not lead to the absurd conclusion that human gametes or somatic cells have serious moral status. Nor do they rely on any of the alleged mistakes in moral reasoning that are made by "potentiality" arguments in ethics. If the arguments of Chapters 2 and 3 are to be turned back, they must be turned back for reasons other than the reasons frequently brought against the Argument From Potential.

Chapter 5
Not Just Damaged Goods: Higher-Order Capacities and the Argument from Marginal Cases

If the arguments of Chapter 2 and 3 are sound, then even the most "marginal" human organisms still have the typical human capacities that are sufficient to generate serious moral status. But this makes my position vulnerable to a version of the "Argument From Marginal Cases" (AMC).

1 The Dreaded Argument from Marginal Cases

The general strategy of the "Argument from Marginal Cases,"[1] as its name suggests, is to argue *from* the moral status of certain "marginal cases" of human beings *to* the moral status of certain non-human animals. The term "marginal" here just means "nonparadigmatic": marginal cases are nonparadigmatic human beings who seem to have (mental) capacities equivalent to the (mental) capacities of non-human animals. The AMC argues that, since this seeming equivalence of (mental) capacities is real, and since any plausible criterion of moral status must be spelled out in terms of (mental) capacities, consistency requires us to conclude that the marginal cases and the non-human animals have an equivalent moral status.[2]

Although this is the general strategy of the AMC, particular versions of it vary. Each version has, at its core, the comparison of certain marginal cases and certain non-human animals. At least four things account for the differences between the versions.

First, the AMC comes in both critical and constructive versions. Tom Regan, who defends the claim that non-human animals possess rights, is an exponent of both versions.[3] His formulation of the critical version runs as follows:

---

[1]The label comes from Narveson (1977, p. 167).

[2]Consequently, the AMC is occasionally called the "Argument for Moral Consistency" because it urges us to be consistent in our evaluation of the moral status of non-human animals and marginal cases. See Dombrowski (1997, p. 24).

[3]Regan (1979, pp. 193, 196), quoted in Dombrowski (1997, pp. 27–28).

R. DiSilvestro, *Human Capacities and Moral Status*, Philosophy and Medicine 108,
DOI 10.1007/978-90-481-8537-5_5, © Springer Science+Business Media B.V. 2010

1. Given certain criteria of the possession of rights, some marginal humans and not just all animals will be excluded from the class of right-holders.
2. However, humans, including those who are marginal, do have rights and so belong in the class of right-holders.
3. Therefore, each and every one of the criteria of which (1) is true must be rejected as setting a requirement for the possession of rights.

Regan's formulation of the constructive version of the AMC is this:

1. Humans, including those who are marginal, have rights and therefore belong in the class of right-holders.
2. However, given the most reasonable criterion of the possession of rights, one that enables us to include marginal humans in the class of right-holders, this same criterion will require us to include some (but not all) animals in this class.
3. Therefore, if we include these marginal humans in the class of right-holders, we must also include some animals in this class.

It is worth noting that premise 2 in the critical version and premise 1 in the constructive version are the same.

The second thing that accounts for the differences between the versions of the AMC is the fact that the AMC comes in both weak and strong versions. The weak version attempts to show that if the marginal cases have rights (or some moral status or other; more on this in a moment), then the non-human animals also have rights. The strong version adds to the weak version some justification for the claim that the marginal cases have rights. Both the critical and the constructive versions of the AMC listed above, as formulated, are weak versions, since each simply assumes (premise 2 in the critical version, premise 1 in the constructive version) that the marginal cases have rights.

This difference between the weak and strong versions allows critics of the AMC to turn the argument on its head. Such a critic can admit that the marginal cases and non-human animals have an equivalent moral status, yet deny that either the non-human animals or the marginal cases have rights. For example, if we begin with the weak version of the argument, which says that if the marginal cases have rights, then the non-human animals also have rights, and if we add the claim that the non-human animals do not have rights, it follows that the marginal cases do not have rights either. And some thinkers are willing to accept this. So then, although many use the weak version of the AMC to establish claims like "vegetarianism is morally obligatory," there is nothing about the logic of the weak version that prevents others from using it to establish claims like "cannibalism is morally permissible."

There is a third thing that accounts for the differences between the versions of the AMC, which has less to do with the argument's form and more to do with its content. Although the AMC is often used by animal rights theorists to refute various criteria for the possession of rights, its overall structure can be used even when the language of rights is not. This is because the comparison at the heart of the AMC is a useful heuristic device for testing any moral status concept. For example, a constructive version of the AMC formulated in terms of serious moral status might run as follows:

1. Humans, including those who are marginal, belong in the class of those who have serious moral status.
2. However, given the most reasonable criterion of the basis of serious moral status, one that enables us to include marginal humans in the class of those who have serious moral status, this same criterion will require us to include some (but not all) animals in this class.
3. Therefore, if we include these marginal humans in the class of those who have serious moral status, we must also include some animals in this class.

The fourth thing that accounts for the differences between the versions of the AMC also has to do with its content. Different versions of the AMC can be generated depending on which marginal cases and which non-human animals are being compared. There is obviously great variety among non-human animals, and there are also many different types of marginal cases. A good illustration of this is found in those passages, quoted in Chapter 1 of this book, from the beginning of Jeff McMahan's book, *The Ethics of Killing: Problems at the Margins of Life*. McMahan claims that there are "four distinct categories into which we may sort most or all instances of killing for which there may be a reasonable justification,"[4] and one of these categories includes "cases in which the metaphysical or moral status of the individual killed is uncertain or controversial." The way he begins his discussion of this category illustrates the heterogeneity of the class of marginal cases:

> Among those beings whose nature arguably entails a moral status inferior to our own are animals, human embryos and fetuses, newborn infants, anencephalic infants, congenitally severely retarded human beings, human beings who have suffered severe brain damage or dementia, and human beings who have become irreversibly comatose.[5]

Some of the marginal cases seem to have (mental) capacities that are not equivalent to, but lower than, many non-human animals. (Indeed, some of the marginal cases seem to have (mental) capacities that are the equivalent of vegetables.) The content of any particular version of the AMC will be a function of the specific sorts of beings compared.

The AMC is a hot topic in contemporary applied ethics. A recent book-length treatment of this argument goes so far as to say that the AMC is "an argument that has generated perhaps more light and heat than any other argument in moral philosophy over the last 20 years."[6] Although it is possible to debate the moral status of marginal cases outside of the context of the AMC, the AMC is often lurking in the background of contemporary debates about marginal cases.

---

[4]McMahan (2002, pp. vii–viii).

[5]McMahan (2002, p. vii).

[6]Dombrowski (1997, p. 3).

My position appears to be vulnerable to a version of the AMC. This vulnerability can be expressed in the form of a dilemma. On the one hand, my position seems to be committed to the idea that many non-human organisms that we know of have a set of typical human capacities, and hence have serious moral status. To see why, focus again on just a single typical human capacity: the capacity to think. Just as the technology of the future might enable us to produce changes in body of a mentally deficient human organism so as to allow her to think like a normal adult human organism, so too the technology of the future might enable us to produce changes in the body of a non-human organism (e.g. a chimpanzee) so as to allow the non-human organism to think like a normal adult human organism. So, if the mentally deficient human organism has a passive higher-order capacity to think right now, on account of what technology might be able to do in the future, the exact same thing can be said about the non-human organism. Of course, the same sort of thing can be said for any typical human capacity. And having the set of typical human capacities is sufficient to generate serious moral status, whether or not the thing having this set is a member of the human species.

On the other hand (and this is the second horn of the dilemma), if my position rejects the idea that many non-human animals that we know of have a set of typical human capacities, and hence have serious moral status, then my position is guilty of some morally objectionable form of "anthropocentrism" or "speciesism". This is because my position attempts to draw an arbitrary metaphysical or moral line between the humans and non-humans, or between our species and other species. Such line-drawing presumably commits one to the same sort of metaphysical and moral arbitrariness that other morally objectionable line-drawing commits one to—such as sexism and racism.

If my position is indeed forced into this dilemma, this would re-emphasize the resilience of the AMC, since the possession of a certain set of typical human capacities turns out to be just the sort of criteria that plugs in nicely to Tom Regan's constructive argument mentioned above. Although an appeal to higher-order capacities does have the benefit of allowing us to say what many want to say about marginal cases, it does this only at the cost of admitting that the non-human animals that we are aware of have serious moral status.

However, I will now argue that my position is not forced into this dilemma. Marginal cases of human organisms can be seen to have serious moral status, without being forced into admitting that the non-human animals we are aware of have serious moral status. And this result can be achieved even without resorting to "speciesist" or "anthropocentric" maneuvering.

2 Tooley's Cat, Boonin's Spider, McMahan's Dog, and Balaam's Ass

Before explaining why my position is not committed to the first horn of the dilemma, it is important to begin by examining the writings of three philosophers who have, in their own ways, formulated arguments which best express the thrust of this first horn: Michael Tooley,[7] David Boonin,[8] and Jeff McMahan.[9]

First, Michael Tooley constructs a thought experiment in which a kitten gets an injection that makes the kitten capable of developing thought patterns just like normal adult human thought patterns. Although this thought experiment is part of a longer complex argument against the moral relevance of potentiality, it is the thought experiment itself (and not the longer complex argument) that is relevant to the first horn of the dilemma:

> Suppose that at some time in the future a chemical is discovered that, when injected into the brain of a kitten, causes it to develop into a cat possessing a brain of the sort possessed by normal adult human beings. Such cats will be able to think, to use language, to make decisions, to envisage a future for themselves, and so on—since they will have all of the psychological capacities possessed by adult humans.[10]

The relationship between Tooley's thought experiment and the first horn of the dilemma is this. My solution to the problem of marginal cases claims that if a future technology could transform an organism so that the organism possesses the immediate capacity to think, it follows that the organism had the higher-order capacity to think to begin with. Tooley's thought experiment simply fills in the details by making the technology an injection and the organism a kitten. Thus my solution to the problem of marginal cases would seem to imply that, in Tooley's thought experiment, the kitten had the higher-order capacity to think to begin with. (And since having this and other typical human capacities at some level or other generates serious moral status, it follows that the kitten, even before it received its injection, would possess serious moral status.)

A more explicit formulation of the first horn of the dilemma is found in David Boonin's discussion of what he calls "the species essence argument." Boonin summarizes a version of this argument taken from Stephen Schwartz:

1. A person is "a being who has the basic inherent capacity for thinking in the broadest sense regardless of how developed or blocked it is."
2. ". . .it is an essential property of every living member of the species *homo sapiens* that it has the capacity to function as a person. . ."
3. "the capacity to function as a person confers on one a right to life. . ."

Therefore,

4. "being a member of *homo sapiens* does ensure that one has a right to life."[11]

---

[7] Tooley (1983, p. 192).

[8] Boonin (2003, pp. 24–25).

[9] McMahan (2002, pp. 302–329).

[10] Tooley (1983, p. 191).

[11] Boonin (2003, pp. 23–24). There is also an additional part of the argument that says, "since every human fetus is a member of *homo sapiens*, it follows that every human fetus has a right to life" (p. 24). But this additional part of the argument is not important for focusing on marginal cases.

The relevant part of Boonin's criticism of this argument is his objection to the second premise:

> The claim that every member of *homo sapiens* has the capacity to function as a person is false. There can, for example, be human fetuses with such severe deformities that they will never develop a brain capable of sustaining thought, or even any brain at all. These are human beings who have not even the capacity for functioning as a person and so are not persons on Schwartz' definition of the term.[12]

Boonin then considers, and quickly rejects, a possible reply that is very similar to my solution to the problem of marginal cases:

> One could, I suppose, characterize such a fetus as a person whose capacity for thought simply happens to be "blocked" by a contingent fact about its head. But then it is difficult to see why we should not also call the spider crawling up my window a person. If he were able to develop a big enough brain, he too would be able to function as a person, so he is simply a person whose capacity is blocked by the fact that he will never have a large enough brain.[13]

The relationship between Boonin's discussion and the first horn of the dilemma is this. My solution to the problem of marginal cases claims that an organism can still possess a higher-order capacity to think even if certain physical conditions prevent (or "block") that capacity from being realized. Boonin invites us to consider how the only conditions that prevent (or "block") a spider from having the immediate capacity to think are certain physical conditions. Thus my solution to the problem of marginal cases would seem to imply that a spider has the higher-order capacity to think. (And since having this and other typical human capacities at some level or other generates serious moral status, it follows that the spider would possess serious moral status.)[14]

Finally, Jeff McMahan constructs a thought experiment in which a dog gets genetic therapy that makes the dog capable of developing thought patterns just like the thought patterns of normal human adults. McMahan uses this thought experiment to make a fairly explicit statement of the first horn of the dilemma. Since McMahan's thought experiment includes a technological element (like Tooley's thought experiment) and is explicitly used to make the point of the first horn of the dilemma (like Boonin's discussion), the remainder of this section will be spent giving McMahan's thought-experiment a careful exposition and reply.

There are two basic ideas behind McMahan's thought experiment. (1) If a human being with a genetically determined cerebral deficit can nevertheless count as having the "intrinsic" potential to develop the cognitive characteristics of a mature human being, in virtue of the fact that the genetic therapy of the future can enable such

---

[12] Boonin (2003, p. 24).

[13] Boonin (2003, p. 24).

[14] Boonin also objects to the second premise because of "human beings who permanently lose their capacity for functioning as a person, such as those whose higher brain regions are irreparably destroyed..." (Boonin, 2003, p. 24). Presumably, any reply to Boonin's objection along the lines of Chapter 2 would be rejected for the same reasons.

individuals to overcome these genetically determined cerebral deficits, then a dog can likewise count as having the "intrinsic" potential to develop the cognitive characteristics of a mature human being. (2) If having this "intrinsic" potential is what determines something's moral status, then there is no difference in the moral status of the dog and the moral status of the impaired human.

To appreciate this thought experiment, it is important to locate it in the flow of McMahan's longer discussion of "whether the [human] fetus's potential can plausibly be regarded as a basis for respect" or more generally "as a basis for moral status."[15] McMahan believes that if the fetus's "potential to become a person" is to be a basis for moral status, then this potential must be grounded in the intrinsic properties of the fetus.[16] McMahan then argues that the fetus's potential to become a person is not grounded in the intrinsic properties of the fetus.

He begins by distinguishing three cases. First, *The Normal Fetus* is "a developed fetus that is in every way normal and healthy". Second, *The Fetus with a Chemical Deficit* is a fetus whose brain is developing normally except that it is deficient in a certain chemical (e.g., a neurotransmitter) without which the fetus will never be a person, because without this chemical, the fetus, even when grown up, will have cognitive capacities that do not surpass the cognitive capacities of a chimpanzee. Third, *The Fetus with Cerebral Deficits* is a fetus whose brain is developing abnormally in that if it continues on its developmental path, the fetus will never be a person. McMahan then asks a pivotal question: "On what basis might it be claimed that these latter two fetuses are potential persons?"[17]

McMahan rejects the answer that "these fetuses are both the sort of entity that normally becomes a person," since

> To point out that these two fetuses are entities of a kind whose normal members tend to become persons is not to show that they have the potential to become persons. It is only to note that *normal* members of the kind have that potential. But *these* two members of the kind are not normal members. And their abnormality is precisely that they lack something that is necessary for them to become persons.[18]

McMahan considers a second answer: "In order for X to have the potential to become a Y, it must be *possible* for X to become a Y." McMahan is willing to admit that this answer allows for The Fetus with a *Chemical* Deficit to count as a potential person. After all, he says, we admit that a seed is a potential plant even when it is not given the water it needs to grow into one. But McMahan is not willing to admit that

[15] McMahan (2002, p. 309).

[16] McMahan uses the term "person" as what some logicians call a phase sortal, like the term "adolescent": an entity can fail to be a person at one time, become a person, and then fail to be a person again. And the *grounding* he has in mind seems to be epistemic: "there must be something about the fetus now that *justifies the claim* that it is a potential person" (McMahan, 2002, p. 309, emphasis mine).

[17] McMahan (2002, p. 310).

[18] McMahan (2002, p. 310).

this answer allows The Fetus with *Cerebral* Deficits to count as a potential person, and he presents two parallel thought experiments to show why.[19]

His first thought-experiment can be summarized as an argument:

(1) A child born without eyes a thousand years ago did not have the potential for sight.
(2) A child born without eyes in a world in which eye transplants are routinely performed would have the potential for sight.
(3) But there is no intrinsic difference between these children.

Therefore,

(4) The difference between the potentials of these children cannot be a matter of their intrinsic properties.

The second thought experiment has the same form as the first:

(1*) A fetus with cerebral deficits in a world, such as ours, in which cerebral augmentation is not possible, does not have the potential to become a person.
(2*) A fetus with cerebral deficits in a world in which cerebral augmentation through genetic therapy is possible would have the potential to become a person.
(3*) But there is no intrinsic difference between the fetus with cerebral deficits in our world and the fetus in the world in which cerebral augmentation is possible.

Therefore,

(4*) The difference between the potentials of these fetuses cannot be a matter of their intrinsic properties.

McMahan characterizes the genetic therapy, in premise (2*), as:

> A form of genetic therapy that, if administered to the fetus with cerebral deficits, would cause it [to] grow the cerebral tissues necessary for normal cognition, and. . .the growth of these tissues would be identity-preserving.[20]

McMahan thinks the upshot of these thought experiments is this. If The Fetus with Cerebral Deficits has the potential to become a person, then this potential is not grounded in the fetus's intrinsic properties. Therefore, in the case of The Fetus with Cerebral Deficits, the fetus's potential to become a person cannot be a basis for the fetus' moral status.

Before going on to explain the rest of McMahan's argument, it is important to pause and notice something. If the analysis of higher-order capacities presented above is correct, then each of these thought experiments begins with a faulty first premise. Surely, the child without eyes does have the potential for sight even in

---

[19] McMahan (2002, p. 311).

[20] McMahan (2002, p. 311).

worlds where this potential never gets realized. A similar remark applies to the fetus with the cerebral deficit. Thus, in the second thought-experiment, McMahan should have argued as follows:

(2*) A fetus with cerebral deficits in a world in which cerebral augmentation through genetic therapy is possible would have the potential to become a person.

(3*) But there is no intrinsic difference between the fetus with cerebral deficits in our world and the fetus in the world in which cerebral augmentation is possible.

Therefore,

~(1*) A fetus with cerebral deficits in a world, such as ours, in which cerebral augmentation is not possible, does have the potential to become a person.

This objection against McMahan's argument relies upon the idea that an entity's potentials depend only upon its intrinsic properties plus the laws of nature, and upon the idea that the worlds in (2*) and (3*) are not just any two possible worlds, but are alternative histories of (or alternative times in) the actual world. McMahan's original argument, on the other hand, seems to assume that an entity's potentials depend upon what is technologically possible at a given time in a given world.

But McMahan has an answer to this type of objection. For his very next move evaluates what he calls a "more radical" view of potential that would block his argument from succeeding. He characterizes this "more radical" view as follows:

> As long as it is *physically* possible for the fetus with cerebral deficits to develop the cognitive capacities that are constitutive of personhood in a way that is identity-preserving, that fetus counts as a potential person...what the potential essentially consists in is an intrinsic receptivity to an identity-preserving transformation into a person. This is a fact about the fetus itself: that it is the sort of thing that can in principle be transformed into a person while continuing to exist.[21]

The thought experiment involving the dog now enters the argument as an objection to this "more radical" view of potential:

> If it is physically possible, through some as-yet-undiscovered form of genetic therapy, to augment a defective fetus's brain in a way that will enhance its future cognitive capacities, it is surely physically possible to achieve the same result in an animal—for example, a dog. If, therefore, we claim that a fetus with cerebral deficits is a potential person on the ground that it is physically possible for its brain to develop in ways that would be identity-preserving and would overcome or repair the deficits, we must concede that a dog is a potential person for the same reason. And if we claim that the fetus's potential to become a person is a basis for moral status (because it is grounded in a suitably intrinsic receptivity to transformation), we must concede that a dog has an equivalent status, other things being equal. Since, however, no one would (or should) accept that dogs are potential persons with a moral status appropriate to their nature as such, we must abandon the broad conception of potential that implies that they are.[22]

---

[21] McMahan (2002, p. 311).

[22] McMahan (2002, p. 312).

McMahan's line of reasoning can be reconstructed as a reductio:

(1) X is a potential person at time t if and only if it is physically possible for X at time t to undergo an identity-preserving transformation into a person [Definition of "potential person"]
(2) It is physically possible for a living thing that is a dog at time t, but not a person at time t, to undergo an identity-preserving transformation into a living thing that is a person at a later time t*. [Assumption]

Therefore,

(3) A living thing that is a dog at time t, but not a person at time t, is a potential person at time t. [From (1) and (2)]
(4) If X is a potential person at time t, then X has moral status at time t. [Assumption]

Therefore,

(5) A living thing that is a dog at time t, but not a person at time t, has moral status at time t. [From (3) and (4)]
(6) But a living thing that is a dog at time t, but not a person at time t, does not have moral status at time t. [Assumption]

Therefore,

(7) Statement (4) is false. [From (1), (2), (5), and (6)]

McMahan concludes that, no matter how the genetic therapy is described, there is always going to be the non-human animal as a counterexample:

> There is, I believe, no basis for claiming that the fetus with cerebral deficits has the potential to become a person that does not also imply that a dog has that potential . . . we should abandon the ambition to include fetuses with cerebral deficits within the category of potential persons.[23]

The relationship between McMahan's discussion and the first horn of the dilemma is this. McMahan's reductio can be easily recast using the concepts of higher-order capacities and serious moral status:

(1a) X has a higher-order capacity to think if and only if it is physically possible for X at time t to undergo an identity-preserving transformation into an entity with the immediate capacity to think [Definition of "higher-order capacity to think"]
(2a) It is physically possible for a living thing that is a dog at time t, but not an entity with the immediate capacity to think at time t, to undergo an identity-preserving transformation into a living thing that is an entity with the immediate capacity to think at a later time t* [Assumption]

---

[23] McMahan (2002, p. 312).

Therefore,

(3a) A living thing that is a dog at time t, but not an entity with the immediate capacity to think at time t, is an entity with a higher-order capacity to think at time t. [From (1a) and (2a)]

(4a) If X is an entity with a higher-order capacity to think at time t, then X has serious moral status at time t. [Assumption]

Therefore,

(5a) A living thing that is a dog at time t, but not an entity with the immediate capacity to think at time t, has serious moral status at time t. [From (3a) and (4a)]

(6a) But a living thing that is a dog at time t, but not an entity with the immediate capacity to think at time t, does not have moral status at time t. [Assumption]

Therefore,

(7a) Statement (4a) is false. [From (1a), (2a), (5a), and (6a)].

In summary, then, the first horn of the dilemma—represented by the arguments of Tooley, Boonin, and McMahan—claims that if certain marginal humans are allowed to count as possessing the higher-order capacity to think, in virtue of what the technology of the future can enable them to do, then certain non-human animals we are aware of must also be allowed to possess the higher-order capacity to think.

The first horn of the dilemma relies upon the assumption that it is possible to perform these transformations on a non-human animal: more precisely, the assumption is that the non-human organisms we are aware of could be given the immediate capacity to think like a normal human adult while still continuing to exist. This assumption is present in the three authors just examined. For example, Tooley's claim that the injection "causes it [the kitten] to develop into a cat possessing a brain of the sort possessed by normal adult human beings"[24] relies upon the assumption that the living thing that is a kitten before the injection is the same living thing as the living thing after the injection. Likewise, Boonin's claim that "If he [the spider crawling up my window] were able to develop a big enough brain, he too would be able to function as a person..."[25] relies upon the assumption that the living thing that is a spider before the increase in brain size is the same living thing as the living thing after the increase in brain size. Finally, McMahan assumes that the living thing that is the dog before the genetic therapy is the same living thing as the living thing after the genetic therapy: "If, therefore, we claim that a fetus with cerebral deficits is a potential person on the ground that it is physically possible for its brain to develop in ways that would be identity-preserving and would overcome or repair the deficits, we must concede that a dog is a potential person for the same reason."[26]

[24] Tooley (1983, p. 191).

[25] Boonin (2003, p. 24).

[26] McMahan (2002, p. 312).

My strategy for replying to the first horn of the dilemma is to deny this assumption. I claim that it is not possible to perform these kinds of transformations on a non-human animal: more precisely, it is simply not true that the non-human animals we are aware of could be given the immediate capacity to think like a normal human adult while still continuing to exist.

This envisaged strategy could be fleshed out in one of two ways. First, it could be making a claim about the limits of technology. It may well be the case that, as a matter of fact, the technology of the future will never be able to (for example) take a dog, produce certain genetic changes in it, and end up with an organism that thinks just like a normal human. This is because the sorts of changes envisioned would be physically impossible, due to the way the bodies of organisms work. If the cells in a dog brain reject or attack the human tissues that are injected into that dog brain, the dog will not be able to develop the immediate capacity to think like a human thinks. I am inclined to think that this first way of fleshing out the strategy is more promising than McMahan is willing to allow. But I really do not know enough about the biology involved in such tissue transplants to say for sure. So I will not pursue this first way of fleshing out the envisaged strategy further.

The second way of fleshing out this strategy is to claim that, even if the tissue injected into the dog was not rejected, still, as a matter of metaphysics, the genetic changes done to the dog would not transform the dog in an identity-preserving way. I would like to explore in a bit more detail the two basic ideas that would need to be accepted to sustain this way of fleshing out this strategy:

(1) All organisms are members of natural kinds, which means, among other things, that certain sorts of changes to a given organism will be identity-preserving changes and other sorts of changes to that organism will be identity-undercutting changes.
(2) The changes required to give the ability to think to any of the non-human animals that we are aware of would be identity-undercutting changes.

Although these ideas are related, they are also importantly distinct. One could accept the first but not the second, and vice-versa.

There are different ways of defending (1). One way is to rely on the claim that biological species are natural kinds. Since all organisms are members of biological species, it would directly follow that all organisms are members of natural kinds. Unfortunately, this way of defending (1) runs into what is sometimes called the "species problem." The proper analysis of the concept of a species is a long-standing debate among biologists and philosophers of biology, and this debate shows no signs of being solved in the foreseeable future. There are at least a dozen different rival accounts of what it means for *x* to be a member of the same species as *y*.[27] While some accounts emphasize the phenotypic similarities between *x* and *y*, others emphasize their genotypic similarities, or reproductive potential, or geographical proximity, or propinquity of descent, or some other common feature and/or

[27] For examples, see following anthologies: Ereshevsky (1992) and Wilson (1999).

relationship. Because of this intractable disagreement, defending (1) using the concept of a species would require being very clear about which concept of a species is in view. It would also require justifying this particular concept of a species rather than the other concepts on offer.

Fortunately, there is an easier way of defending (1) that avoids the concept of a species entirely. This is to rely on the claim that there is some unique way of constructing a biological taxonomy such that the most basic categories of this taxonomy are natural kinds. Since all organisms fall into the categories of this biological taxonomy, it would directly follow that all organisms are members of natural kinds.

Many philosophers think that the members of a natural kind possess something intrinsic in common with each other. Elliot Sober, for example, after stating that "a standard philosophical view about natural kinds" is essentialism, which "holds that each natural kind can be defined in terms of properties that are possessed by all and only the members of that kind,"[28] goes on to explain that an essentialist definition of gold "must cite a property that is intrinsic to gold things; the cited property [in this case, atomic number 79] does not require that any relations obtain among gold things." Likewise, T. E. Wilkerson claims that one of the conditions that must be met by any interesting account of natural kinds is that "members of natural kinds have real essences, intrinsic properties that make them members of the relevant kind, and without which they could not be members of the relevant kind."[29]

If humans and non-human animals are members of different natural kinds, then it follows that humans do share something special in common with each other that they do not share with non-human animals: namely, a real essence. McMahan's argument is looking for an intrinsic difference between dogs and humans; the idea that humans and dogs are natural kinds presents an intrinsic property that a human being possesses and a dog does not. Human beings share a real essence; dogs share a real essence; these real essences are different from each other, but each real essence is intrinsic to the respective organisms that possess it.

However, in order for a doctrine of natural kinds to adequately answer the first horn of the dilemma, it needs to be careful not to identify an organism's real essence with its genome. For consider again the fetus with cerebral deficits. If her set of potentials or dispositional properties is identified with her real essence, and if her real essence is identified with her genome, then it follows that she is not a potential person. After all, her genetic material was the problem to begin with, since it seemed to undercut the idea that she is a potential person.

The contemporary approach to natural kinds, as represented by Kripke and Putnam, usually attempts to identify a thing's real essence with some structural feature of that thing. For example, if something looks like gold but does not have atomic number 79, then it is not gold; likewise, if something is functionally just like water but does not consist of H_2O, then it is not water. When this understanding of natural kinds is introduced to do work in constructing a biological taxonomy, it will end up emphasizing a genetic real essence. For example, if something looks like a

[28] Sober (1993, p. 145).

[29] Wilkerson (1988, p. 29).

duck, walks like a duck, and quacks like a duck, this still does not guarantee it is a duck. It needs to possess the genetic real essence (whatever precisely it is) that all real ducks have. Likewise, if "human" is a natural kind like "gold" or "water", then unless something possesses the genetic real essence (whatever precisely it is) that all real humans have, that thing will not be a human.

This is a problem that even an advocate of biological natural kinds must confront. For example, T. E. Wilkerson argues that natural kinds exist in biology, but that species are not natural kinds. Wilkerson considers an objection to a Kripke-Putnam approach, namely, the objection that "species are not uniquely determined by genetic constitution" or, in other words, "genetic real essences of natural kinds do not exist."[30] This objection to a Kripke-Putnam approach can point to various examples in real-life biology: there may be a good deal of genetic variation between the parts of the same individual (in plants) or between the members of the same biological kind; conversely, there may be a good deal of genetic similarity between closely related species.[31]

One reply to this objection might be to claim that: "the genetic feature we are looking for is a structural feature of the genetic material—for example, the number of chromosomes peculiar to each species."[32] But Wilkerson rejects this reply, since not all humans have 23 pairs of chromosomes (for example, those with Down's syndrome have an extra chromosome) and many plants have multiple sets of chromosomes (a feature called polyploidy). Wilkerson's conclusion is that "the more we attempt to isolate the genetic features that determine biological species, the more hopeless the task becomes."[33] The problem, as he restates it, is this:

> If natural kinds are determined by real essences, and if species are good examples of natural kinds, then we appear to have produced a contradiction, since species are not determined by real essences.[34]

An obvious solution to this problem is to claim that species are determined by real essences, but that these real essences are not genetic. Wilkerson's own solution, however, is to keep the real essences genetic and to simply increase the number of natural kinds in biology. Here is how he summarizes it:

> There are natural kinds. Each natural kind is determined by a real essence, a property or set of properties necessary and sufficient for membership of the kind in question. The real essence in turn grounds the causal powers of individual members of the kind. Biological natural kinds are determined by genetic real essences which are causally responsible for the behaviour of individual members of the kind. But, since there is considerable interspecific genetic similarity and intraspecific genetic variation, there are far more biological natural kinds than species.[35]

---

[30] Wilkerson (1993, p. 7).

[31] Wilkerson (1993, pp. 7–8).

[32] Wilkerson (1993, p. 8).

[33] Wilkerson (1993, p. 8).

[34] Wilkerson (1993, p. 10).

[35] Wilkerson (1993, p. 16).

Whatever the merits of Wilkerson's proposal may be for systems of biological taxonomy, his proposal is not helpful for explaining how a human with a genetic handicap can be a member of the same natural kind as you and I. For if the real essence that grounds membership in biological natural kinds is equated with the genome, then there would be as many natural kinds as there are genetically diverse individuals. But this proliferation of natural kinds is tantamount to giving up on the doctrine of natural kinds in biology. Imagine a "biological periodical table", if you will, that is designed to parallel the periodic table of the elements. Just as gold has its own box with its own atomic number, so each genetically distinct individual would have its own box with its own (let us say) genetic number. If a population of genetically identical clones existed, then of course many different instances of the genetic real essence would exist. But everyone else would be all by himself or herself in the biological periodic table. Consequently, if one wants to explain why the human organism with a cerebral deficit and the dog are members of different natural kinds, Wilkerson's proposal works just fine. But if one wants to explain why the human organism with a cerebral deficit and the normal human organism are members of the same natural kind, the real essence they share must go deeper than, and be different from, a purely genetic constitution. Otherwise, the human with a genetically determined cerebral deficit will be a member of a different natural kind from the rest of us.

There are at least three reasons for believing that an organism's real essence is deeper than, and different from, its genetic constitution. First, the possibility of an organism surviving a small genetic mutation suggests that an organism's genetic code should not be equated with its real essence. Second, the fact that it is possible to have more than one genome within the same organism also suggests that an organism's genetic code should not be equated with that organism's real essence. Third, the following thought experiment gives us a reason for not identifying an essence with a genotype. Suppose you have a one-celled organism. On Monday, you put it into a state of suspended animation and extract its DNA. For the next 3 days, you tinker with its DNA in a separate part of the lab. On Friday, you reinsert the DNA back into the organism and thaw out the organism. The organism exists even when its DNA has been removed. In other words, the organism continues to have its essence even when its genotype has been removed. Therefore its essence is not its genotype.

One might object to the plausibility of describing this case the way I have described it on the following grounds: "it seems like the one-celled organism is killed, and then a new one-celled organism, with the same DNA as the first organism, is brought into being—a process somewhat like cloning. More generally, genotype seems *necessary* for something to be an organism at all (even if genotype is not *sufficient* for something to be an organism). The entity with no DNA (in this example) has no self-actualizing capacities at all."[36]

---

[36]Many thanks to Chris Tollefsen for pressing me on this point.

My response to this objection moves in two steps. First, I believe that the pretheoretical language we would use in describing this case is suggestive of the interpretation I have already given. For example, we would say that the organism gets "its" (the organism's) DNA removed and then returned to "it" (the organism). Second, I believe this first point is at its strongest whenever the extraction-alteration-reinsertion of the DNA is made with an eye towards *benefitting* the organism. If the DNA were extracted from a one-celled elephant in order to be tinkered with for the sake of that very elephant (for example, say the scientists are genetically modifying the genes that code for the functionality of Dumbo's elephant trunk, so that when Dumbo grows up he can use his trunk like all the other normal elephants), I believe we would be strongly and rightly tempted to see this as a therapeutic intervention, which begins with Dumbo, ends with Dumbo, and has Dumbo in the middle. The same thing would apply, I believe, if the genetic "therapy" was being attempted on a one-celled human embryo.[37]

The essence of an organism is partly characterized, but not fully exhausted, by making reference to phenotypes and genotypes. The essence is phenotypic in the sense that the essence is characterized by the range of phenotypes an organism can exhibit while still remaining the same organism. But the essence is not *strictly* phenotypic, because the characterization referencing the phenotypes is subjunctive: the important thing is not what phenotypes the organism actually has, at this very moment, but rather what phenotypes it *would* have if it were put in this or that set of circumstances. Consequently, an organism can still have its essence before any of its possible phenotypes have been expressed, and the capacity to realize a given phenotype is more central to its essence than the actual expression of that phenotype. The essence is genotypic in the sense that a genetic structure of an organism is often part of the physical basis for an active higher-order capacity. But the essence is not *strictly* genotypic, because it is not to be identified with the genetic structure of an organism.

Idea (2), the second basic idea behind this strategy, was that the changes required to give the ability to think to any of the non-human animals that we are aware of would be *identity-undercutting* changes. To grasp what this means, consider first the concepts of incapacities and essential incapacities. An incapacity is simply an inability to do something: for example, I have an incapacity for speaking Chinese, an incapacity for remembering your experiences, and an incapacity for omnipresence, omniscience, and omnipotence. An essential incapacity is an essential inability to do something, an inability that one cannot lose while remaining who one is: for example, I have an essential incapacity for remembering your experiences, since if

[37] This response leaves open the question of which genetic interventions are genuinely therapeutic, and which are merely frivolous (or harmful) even though they were intended to be therapeutic. If the scientists are genetically modifying the genes that code for the size of Dumbo's elephant ears, so that when Dumbo grows up he looks just like all the other normal elephants, this may prevent Dumbo from having various positive and negative experiences that he would have *precisely because of* his oversized ears. Similar things might be said about parallel cases using human embryos.

I were to lose this incapacity, I would become someone else: namely, you. Likewise, I have an essential incapacity for omnipresence, omniscience, and omnipotence, since if I were to lose this incapacity, I would become someone else: namely, God. However, my incapacity for speaking Chinese is not an essential incapacity, since I can lose it while remaining who I am.

Philosophers have sometimes claimed that a property can be essential to an object in one of two ways. First, a property is *kind-essential* if its being had by an individual is needed for that individual to belong to a particular kind. Second, a property is *individually-essential* if the individual that has it could not have existed without having it. Some philosophers have thought that, if a given property is kind-essential to an individual, then that property is also individually-essential to that individual. Others prefer to make room for the thought that a given property might be kind-essential to an individual without being individually-essential to that individual.

Idea (2) should be read as the claim that each of the particular non-human animals we are aware of has an individually-essential incapacity to obtain the immediate capacity to think. As before, however, let me emphasize that I focus on the capacity to think just for the sake of convenience. What matters is that each of the non-human animals we are aware of has an individually-essential incapacity to obtain the whole set of typical human immediate capacities, one of which is thinking.

Another way of grasping the idea behind (2) is to invoke the concept of a *modal boundary*. There are innumerable ways that an entity can be modified, but an entity's modal boundary is the metaphysical line beyond which that entity cannot go. For example, the caterpillars we are aware of would cross their modal boundary if they changed into puppies. But the caterpillars we are aware of do not cross their modal boundary merely by changing into butterflies. On the other hand, if we discovered a group of organisms that looked like caterpillars, but that changed into puppies, we would not say that these organisms were caterpillars that had crossed their modal boundaries. We would say that they were not caterpillars at all: perhaps we would call them "scatterpillars". Even the character from Greek mythology named Proteus had his modal boundaries: even though he could take on the typical capacities of a donkey, and then take on the typical capacities of a human, and so on, still, he could not become omnipresent, omniscient, and omnipotent.

One of the best reasons for believing in basic idea (2) comes from attempting to construct a temporary change argument in which a human temporarily changes into a non-human animal. It seems that there are certain sorts of apparent temporary changes that are not temporary changes at all, but are rather instances of one individual ceasing to exist and another individual coming to exist. Recall the example above where a scientist accidentally steps in front of a machine while it is emitting A-rays. Imagine that the administration of A-rays, instead of transforming this scientist into an individual with a genetic disability, instead transforms the scientist into a dog. I believe that the scientist before the administration of A-rays is *not* the same organism as the dog after the A-rays. The apparent transformation of the scientist was actually not a case of transformation at all, but one of annihilation and creation: the original scientist was annihilated (or disembodied, for those who believe in the possibility of disembodied existence), and a brand new organism was created. And this is so *even if* the machine is able to emit B-rays that "transform" the dog into a human organism (who we may call "the resultant scientist") a few minutes later. The B-rays would not actually transform the dog, but would annihilate (or disembody) it and create the resultant scientist in its place. The resultant scientist would not be the original scientist.

The hypothetical example of "scatterpillars" tells us something important. If we came across a particular non-human animal that appears to have been transformed, in an identity-preserving way, so that it now seems to have the typical human capacities, such as thinking, we have two options for describing this. On the one hand, we could say that this apparently identity-preserving transformation was actually identity-undercutting, and that the original animal ceased to exist (or at least ceased to exist *right there*, if one wishes to make room for disembodied non-human animal souls) at the moment a new organism began to exist. On the other hand, we could say that this apparently identity-preserving transformation was indeed genuinely identity-preserving, and that we were mistaken in our original thought that this particular non-human animal possessed an individually-essential incapacity to obtain the typical human capacities. The individual that changed was a "scatterpillar" after all, and not a caterpillar.

Denying the numerical identity of the original dog and the resultant organism with the immediate capacity to think is simply the third step in a natural progression, the first two steps of which have already been seen to be acceptable. The first step in this progression (from Chapter 4) dealt with re-arranging an island of raw materials so that it came to constitute a human organism. The mere facts that the original individual (the island) was spatiotemporally continuous with and mereologically indistinguishable from the resultant individual (the human organism) were not sufficient for making the original individual numerically identical to the resultant individual.

The second step in the progression (also from Chapter 4) dealt with changing a part of a human organism so that it came to constitute a new human organism. Here, as in the first step, the original individual was spatiotemporally continuous with and mereologically indistinguishable from the resultant individual. But in addition, the original individual and the resultant individual were both alive and were genetically identical to one another (i.e. they had the same genetic code). Yet these facts were not sufficient for making the original individual numerically identical to the resultant individual.

Whether one is a materialist or a dualist of one stripe or another, one of the lessons learned from these first two steps is this: *whatever it is* that provides the locus of identity through time for an organism, it is not the elements of spatiotemporal continuity or mereological indiscernability. Nor is it the combination of these elements under conditions where both the original individual and the resultant individual are alive. Therefore, one cannot rely upon these elements when one claims that a dog before the application of technology is the same individual as the resultant individual after the application of technology.

There are at least three objections to this strategy and to the claims that back it up. First, it might be objected that any doctrine of natural kinds in biology is inconsistent with the theory of evolution. In reply, this objection is mistaken for a reason Elliot Sober gives in his evaluation of different evolutionary arguments against the idea that *species* are natural kinds. Even though the above discussion avoided the species concept, and even though Sober himself does not think species are natural kinds, relying upon him here is instructive because the point he makes is relevant to evolutionary arguments against essentialism in biology. One such argument runs as follows: (1) natural kinds are immutable; (2) species evolve; therefore (3) species are not natural kinds. Sober replies that, just as an atom smasher can transform lead into gold without undermining the idea that the chemical elements have immutable essences, "the fact that a population belonging to one species can give rise to a population belonging to another species does not refute essentialism about species."[38] And just a page later, he says:

> In general, essentialism is a doctrine that is compatible with certain sorts of vagueness. The essentialist holds that the essence of gold is its atomic number. Essentialism would not be thrown into doubt if there were stages in the process of transmuting lead into gold in which it is indeterminate whether the sample undergoing the process belongs to one element or to the other. I suspect that no scientific concept is *absolutely* precise; that is, for every concept, a situation can be described in which the concept's application is indeterminate. Essentialism can tolerate imprecisions of this sort.[39]

Sober's point about the transmutation of elements is relevant to the mutation of organisms. It seems that essentialism about chemical elements does not stand or fall depending on one's theory about the origin and historical evolution of these elements. For example, if it were to be discovered that all of our present elements originally emerged, very gradually, from a sort of primordial stuff, this would not throw into doubt the theory of natural kinds about gold. But then why should it be any different for biological organisms? It would seem that essentialism about biological organisms does not stand or fall depending on one's theory about the origin and historical evolution of these organisms. For example, if it were to be discovered that all of our present organisms originally emerged, very gradually, from a sort of primordial soup, this should not throw into doubt the theory of natural kinds about humans.

A second sort of objection is that this strategy is just very difficult to believe, because it does seem possible to imagine a non-human organism being transformed, in an identity-preserving way, in such a way that it comes to think just like a human being. Such transformations pop up all the time in various kinds of literature around the world. Consider the Biblical story of Balaam's donkey:

> Balaam got up in the morning, saddled his donkey and went with the princes of Moab. But God was very angry when he went, and the angel of the LORD stood in the road to oppose him. Balaam was riding on his donkey, and his two servants were with him. When the donkey saw the angel of the LORD standing in the road with a drawn sword in his hand, she turned off the road into a field. Balaam beat her to get her back on the road. Then the angel of the LORD stood in a narrow path between two vineyards, with walls on both sides. When

[38] Sober (1993, pp. 146–147).

[39] Sober (1993, p. 148).

> the donkey saw the angel of the LORD, she pressed close to the wall, crushing Balaam's foot against it. So he beat her again. Then the angel of the LORD moved on ahead and stood in a narrow place where there was no room left. When the donkey saw the angel of the LORD, she lay down under Balaam, and he was angry and beat her with his staff. Then the LORD opened the donkey's mouth, and she said to Balaam, "What have I done to you to make you beat me these three times?" Balaam answered the donkey, "You have made a fool out of me! If I had a sword in my hand, I would kill you right now." The donkey said to Balaam, "Am I not your own donkey, which you have always ridden, to this day? Have I been in the habit of doing this to you?" "No," he said."[40]

Whether or not the event described in this story was historical, it is surely metaphysically possible. And yet this event seems to embody precisely the sort of identity-preserving transformation (of the donkey) that the above view of natural kinds says is not metaphysically possible. The organism before the divine intervention and the organism after the divine intervention are the same organism: "Am I not your own donkey, which you have always ridden to this day?"

One way to reply to this objection is to say that, although the events in these stories are intelligible, the proper interpretation of these events need not admit that the individuals in these stories retained their identity through time. Of course, it may have *seemed* to Balaam, and his donkey, that the organism after the divine activity was the same individual as the donkey before the divine activity. After all, the resultant organism had at least apparent memories of being beaten, and supposedly it had similar mental states to the mental states of the original donkey. But these sorts of considerations—apparent memories, similar mental states—are notoriously insufficient for genuine identity through time. Perhaps it is preferable to say that Balaam and his donkey were mistaken.

But another way to reply to this objection is to say that, *in this particular case*, the donkey did not have an individually-essential incapacity to think. Saying this does not commit one to the view that *all* donkeys have a passive higher-order capacity to think. It only commits one to the view that Balaam's donkey had a passive higher-order capacity to think. And if it turned out that Balaam's donkey had a set of higher-order typical human capacities, then, in the case of Balaam's donkey, we would have to admit that it had serious moral status. When confronted with the case of Balaam's donkey, we employ the same move that we employ when confronted with a scatterpillar. In both cases, we simply admit that the individual substance was not the sort of thing we had assumed it was at the beginning, and that it did not have the modal boundary that we had originally assumed it had.

A third objection to this strategy agrees with basic idea (1) but denies basic idea (2). There are various alternatives for individuating natural substances. For example, perhaps the relevant substantial kind that a spider is a member of is not *spider* but rather *living animal*. On this view, as long as a living animal that is a spider remains a living animal, then that living animal will still continue to exist no matter what sorts of changes it undergoes: the living animal that is a spider could receive a huge brain, and learn philosophy, and indeed could even become a dog or a cat or a

[40]Numbers 22:21-30 NIV (New International Version).

human. Or perhaps the relevant substantial kind that a spider is a member of is not *living animal,* but *living organism.* On this view, a living organism that is a spider could become an oak tree or an amoeba while still continuing to exist.

The best reply to this objection is that choosing from among these alternatives for the relevant substantial kind—living animal, living organism, and so on—is fundamentally and inevitably a matter of testing our modal intuitions with thought experiments of various kinds. Each alternative has its interesting results, or philosophical bullets to bite, and the task of choosing between them is one of deciding which bullets are the least worst to bite. I believe that the alternative I have suggested is at least as good as, and indeed, better than, the ones suggested by this objection. Does it really seem plausible to claim that the same living organism that is now a spider could one day become an oak tree, an amoeba, a human, and a dog?[41]

3 How Not to Be a Speciesist

The second horn of the dilemma claims that my position is "anthropocentric" or "speciesist" in some morally objectionable way. But this objection is mistaken, and several of the reasons it is mistaken have already been touched upon in one way or another in the discussion of the first horn of the dilemma. For example, the strategy discussed above deliberately avoided the concept of a biological species when defending basic idea (1). The discussion of real essences and genetic codes was neutral between humans and non-human animals. Likewise, when defending basic idea (2), the idea of a temporary change was revised to take account of the fact that humans, too, can only undergo certain sorts of changes and retain their identity through time. Certain kinds of identity-undercutting changes can happen to living human organisms.

---

[41] One reviewer asked whether my treatment of this topic would still be the same if the envisaged modification to the non-human animal were not a modification involving any changes to the genetic structure of the non-human animal. My answer, perhaps not surprisingly, is yes. If an entity has a given modal boundary, that boundary may not be crossed, period. If an entity has a given essential incapacity, that incapacity cannot be circumvented merely by making non-genetic enhancements rather than genetic enhancements. However, let me take this occasion to propose one other possible way of handling the problem of this chapter that I have not yet developed, much less published, elsewhere. Even if we were to admit that it is possible to take any non-human organism (animal, plant, amoeba, whatever), and change it in identity-preserving ways so that it looks and acts just like a human being does nowadays, and even if we were to thereby admit that this non-human organism possesses, right now, the higher-order capacities (at some very high level) to do the sorts of activities that humans now do (thinking, loving, etc.), this by itself would still not mean that humans have the same sets of higher-order capacities as non-human organisms. For it could be that the set of typical human capacities H of what human are *in fact* able to do is only a sub-set of a larger set H* of capacities of what humans are *in principle* able to do. Call H the *empirical* set of human capacities, and H* the *total* set of human capacities. Likewise, call D the empirical set of dog capacities, and D* the total set of dog capacities. Even if we admit that D* includes H, this does not mean that D* and H* are the same. But this proposal would require much more development.

There are at least four more reasons why my position is not anthropocentric or speciesist in any morally objectionable sense. First, my position does not say that the non-human organisms we are aware of do *not* have serious moral status. My position is a sufficiency account. It is therefore fully compatible with the claim that the non-human animals we are aware of do have serious moral status. It is even compatible with the claim that the serious moral status of non-human organisms is based on their typical capacities. Perhaps the set of typical dog capacities generates serious moral status for anything that possesses it.

Second, my position is willing to admit that, after certain identity-undercutting changes have taken place in a non-human organism, the resulting organism now possesses the set of typical human capacities which generates serious moral status. These resultant organisms have serious moral status, whether or not the original organisms did. For example, if the organism that results from altering Tooley's kitten (or the organism that results from altering McMahan's dog, or the organism that results from altering Boonin's spider) has a set of typical human capacities, then this resulting organism does indeed have serious moral status.

Third, if there are non-human organisms with exactly the same higher-order capacities as humans—for example, if we discovered a race of "shumans" on Mars whose members possess the same set of typical human capacities as you or I, even though they had ZNA instead of DNA in their cells—then these non-human organisms would have serious moral status, even though they were not part of our species. My position is that whatever possesses the set of typical human capacities, whether it is part of our species or not, has serious moral status.

Finally, if some of the non-human animals we are aware of do, in fact, have serious moral status because of the set of capacities they possess, the strategy I used with humans would be quite useful in defending the moral status of the marginal cases of such non-human animals. For just as marginal humans still have the distinctively human set of capacities, so too the marginal chimpanzees still have the distinctively chimpanzee set of capacities. If it turns out that normal adult chimpanzees have serious moral status because of their capacities, then marginal chimpanzees—those suffering from temporary changes, brain damage, mental retardation, and so on—will also have serious moral status. Focusing on the passive higher-order capacities of an organism allows a dog lover to defend the moral status of defective dogs, a cat lover to defend the moral status of defective cats, and a chimpanzee lover to defend the moral status of defective chimpanzees. But it does this without the unwelcome implication that spiders, ticks, and cockroaches have the same moral status as dogs, cats, and chimpanzess.

Chapter 6
Old Objections and New Directions: Capacities and Moral Status at the Very Borders of Human Life

In this final chapter, I would like to consider several possible objections to the view defended in the previous pages. Some of these objections will look somewhat familiar: in a sense, they are old objections, or at least new versions of old objections. Some of these objections are somewhat new, and require extending or applying previous claims and arguments to different situations. Some objections relate to my methodology. Some objections relate to my treatment of specific cases. Some objections relate to both. My treatment here will attempt to begin with a methodological objection, and will progress into specific cases to emerge as illustrations of the methodology.

1 Does the Temporary Change Argument Prove Too Much?

Throughout this book, two closely related types of "metaphysical" temporary change argument have been used:

- If you could temporarily become an X, then you could have been an X.
- If you could temporarily become an X, then you could have been a Y relevantly similar to an X.

For examples of the first type of metaphysical temporary change argument:

- If you could temporarily become a human embryo, then you could have been a human embryo.
- If you could temporarily become a human organism with merely a 100th-order capacity to think, then you could have been a human organism with merely a 100th-order capacity to think.

For examples of the second type of metaphysical temporary change argument:

- If you could temporarily become an infantilized adult, then you could have been a human infant.

R. DiSilvestro, *Human Capacities and Moral Status*, Philosophy and Medicine 108,
DOI 10.1007/978-90-481-8537-5_6, © Springer Science+Business Media B.V. 2010

- If you could temporarily become the capacity-equivalent of a genetically deformed anencephalic human infant, then you could have been a genetically deformed anencephalic human infant.

However, it might be objected that each type of metaphysical temporary change argument is too permissive. For consider parodies of the first type of metaphysical temporary change argument:

- If you could temporarily become a human somatic cell, then you could have been a human somatic cell.
- If you could temporarily become a spatially separated sperm and egg, then you could have been a spatially separated sperm and egg.

Likewise, consider parodies of the second type of metaphysical temporary change argument:

- If you could temporarily become a human corpse, then you could have been a non-living hunk of matter.
- If you could temporarily become the capacity-equivalent of a non-human animal, then you could have been a non-human animal.

In short, a common objection to the methodology of this book is that, if we take the methodology seriously, it leads to absurdity. In each of the parodies above, we can *imagine* you becoming the relevant thing. If this means you could have *been* the relevant thing, then you could have been anything. The listed parodies are only the tip of the iceberg. After all, we can imagine all kinds of temporary change scenarios, in which you temporarily become a rock, and/or a cockroach, and/or a heap of ashes, before returning to your present state. All bets are off once we permit temporary change scenarios in the door.

Even worse (this common objection continues), the methodology of this book uses the results of these metaphysical temporary change arguments to advance theses about moral status. The methodology does this by employing a "moral" temporary change argument that builds upon the "metaphysical" temporary change argument in roughly the following way:

- If you could temporarily become an X, while retaining your serious moral status, then you could have been an X with serious moral status. To put the conclusion differently: Xs have serious moral status, just like you do now.

For example,

- If you could temporarily become a human embryo, while retaining your serious moral status, then you could have been a human embryo with serious moral status. To put the conclusion differently: human embryos have serious moral status, just like you do now.

But this "moral" temporary change argument (it is objected) can be parodied:

- If you could temporarily become a human somatic cell, while retaining your serious moral status, then you could have been a human somatic cell with serious moral status. To put the conclusion differently: human somatic cells have serious moral status, just like you do now.

And likewise if we replace "human somatic cell" with "rock" or "cockroach" or "heap of ashes". But do we really want to say that a human somatic cell, or a rock, or a cockroach, or a heap of ashes, has the same serious moral status as you?

Although I have already addressed this objection at various points in the previous pages, let me now restate, and expand upon, the main lines of my response. The temporary change argument is not a blank check that can be filled in with any metaphysical thesis whatsoever. Rather, the temporary change argument is a heuristic device for testing, sharpening, and challenging our modal and moral intuitions. There are ways out of a temporary change argument, just as there are ways out of the arguments in Chapter 1 concerning our hypothetical character Ben (who thought that a person could not survive the period from age 20 to 30) and our imaginary scenario involving Ronald Reagan.

For example, recall the argument, and some of the ways out, involving the hypothetical case of Ronald Reagan. The argument ran as follows:

(1) Reagan exists before the brain disease does its damage.
(2) Reagan cannot have temporal gaps.
(3) Reagan exists after the injection-plus-scan-in.

Therefore,

(4) Reagan cannot cease to exist when the brain disease does its damage.

One strategy for avoiding the conclusion is to deny (2): perhaps Reagan *can* have temporal gaps. Another strategy for avoiding the conclusion is to deny (3): perhaps the individual that exists after the injection-plus-scan-in is *not* Reagan. However, in the case of Reagan, neither of these conclusion-avoidance strategies is very appealing. We then seem to be forced into accepting the relevant conclusion—that Reagan cannot cease to exist when the brain disease does its damage. If so, then it looks like Reagan can survive this brain disease—as the damaged human organism. And, given the view that Reagan retains his serious moral status throughout this temporary change, the upshot is that he has serious moral status as a brain-damaged human organism. Generalizing from this one case, we arrive at the rough conclusion that all brain-damaged human organisms (at least of a certain sort) have serious moral status.

Now consider a different example, discussed toward the end of Chapter 5: a scientist steps in front of a machine emitting A-rays, and is "transformed" into a dog, and then the machine emits B-rays and "transforms" the dog back into a human organism. Although I didn't say so then, let's assume now for the sake of argument that the human organism that walks away from this adventure (who I called "the resultant scientist" in Chapter 5) is physically and functionally indistinguishable from the human organism that walked into this adventure (who I called "the original scientist" in Chapter 5). Calling the original scientist Sally, one is now in a position to set up a parallel argument to the one about the hypothetical case of Ronald Reagan:

(1) Sally exists before the machine emits A-rays.
(2) Sally cannot have temporal gaps.
(3) Sally exists after the machine emits B-rays.

Therefore,

(4) Sally cannot cease to exist when the brain disease does its damage.

One strategy for avoiding the conclusion is to deny (2): perhaps Sally *can* have temporal gaps. Another strategy for avoiding the conclusion is to deny (3): perhaps the individual that exists after the machine emits B-rays is *not* Sally. However, it might be claimed that in the case of Sally, neither of these conclusion-avoidance strategies is very appealing. We then seem to be forced into accepting the relevant conclusion—that Sally cannot cease to exist when the machine emits A-rays. If so, then it looks like Sally can survive the emission of A-rays—as the dog! And, given the view that Sally retains her serious moral status throughout this temporary change, the upshot is that she has serious moral status as a dog. Generalizing from this one case, we arrive at the conclusion that all dogs (at least of a certain sort) have serious moral status.

Of course, that is not the option I took in Chapter 5. I tried to find a way out. I called the eventual human organism "the resultant scientist," and I claimed that the original scientist was not the resultant scientist. In other words, I denied step (3) in the argument ("Sally exists after the machine emits B-rays"). I claimed that the individual that exists after the machine emits B-rays is *not* Sally. And I would take the same sort of approach in cases where a human organism is supposedly "transformed" into a human somatic cell, or a spatially separated sperm and egg, or a rock, or a cockroach, or a heap of ashes, and then "transformed" back into a human organism.

My reply up to this point may still leave some readers with a serious concern. "Granted," they may say, "there are ways out of a temporary change argument. Granted, the temporary change argument is a heuristic device. But this just means that not all *apparent* temporary changes are *genuine* temporary changes. This, in turn, forces us to ask, and answer, the question: how can we tell, in any given case, whether an *apparent* temporary change is a *genuine* temporary change? The apparent temporary change where Reagan appears to change into a brain-damaged human organism and then back to his normal self is a genuine temporary change: Reagan literally becomes a brain-damaged human organism. But the apparent temporary change where Sally appears to change into a dog and then back to her normal self is *not* a genuine temporary change: Sally does not literally become a dog. But why the different treatment in these two cases?"

There are several ways one could go in answering this concern. One could appeal to intuition: we can (or should) simply see that some apparent temporary changes are genuine, while others are not. Or, one could appeal to a metaphysical theory of persistence through time: a given theory will dictate which apparent temporary changes are genuine and which are not. Or, one could appeal to experience: I know that some apparent temporary changes are genuine, because I've experienced them myself (in the first-person case), or I've seen others go through them (in the third-person case). My own preferred method, as the previous parts of this book illustrate, is to aim for an optimal reflective equilibrium between our intuitions, our metaphysical theories, and our experiences (both first- and third-person). There is some unavoidable give-and-take in trying to achieve such equilibrium: accepting certain intuitions or experiences requires rejecting or adjusting a given theory, and accepting a given theory requires rejecting certain intuitions or experiences.[1]

To illustrate how this works in a concrete case, I would like to consider more closely the objection that my methodology commits me to viewing dead human organisms—human cadavers or corpses—as having serious moral status. Then I would like to apply what we learn from this case more broadly to other contested cases.

2 The Corpse Problem

Consider an argument that attempts to parallel the argument about Ronald Reagan reviewed a few paragraphs ago. Consider also the corresponding escape strategies. The case is one where a human organism apparently dies, and is then somehow revived. Let's say you are a patient who undergoes cardiac arrest, where your heart completely stops beating. Let's say that after the cardiac arrest, you are apparently dead for three minutes. Let's say that you are then revived using cardio-pulmonary resuscitation, or CPR. An argument might go as follows:

(1) You exist before the cardiac arrest.
(2) You cannot have temporal gaps.
(3) You exist after the CPR.

Therefore,

(4) You cannot cease to exist when the cardiac arrest occurs.

As in the parallel argument concerning Reagan, one strategy for avoiding the conclusion is to deny (2): perhaps you *can* have temporal gaps. Another strategy for avoiding the conclusion is to deny (3): perhaps the individual that exists after the CPR is *not* you. However, as in the case of Reagan, neither of these conclusion-avoidance strategies is very appealing. Are we then forced into accepting the relevant conclusion—that you cannot cease to exist when the cardiac arrest

---

[1] See Beauchamp and Childress (2008, Chapter 10).

occurs? If so, then it looks like you can survive your own death—as a cadaver or corpse. And, given the view that you retain your serious moral status throughout this temporary change, the upshot is that your corpse has serious moral status. Generalizing from this one case, we arrive at the conclusion that all corpses (at least of a certain sort) have serious moral status. Let us therefore call this sort of problem "the corpse problem."

One strategy for trying to block this argument is to claim that, *in the case of cardiac arrest just described*, you were never dead to begin with. Hence your "survival" between the time of cardiac arrest and the time of CPR is not the survival of your own death. Hence your continued possession of serious moral status during this time interval does not have any implications about the moral status of corpses. Let us consider this strategy more closely.

There are genuine medical and philosophical controversies regarding when a human organism has died. One of the most recent attempts to get clear on these controversies and make progress in resolving them is a January 2009 report by the President's Council on Bioethics titled "Controversies in the Determination of Death."[2] This report illustrates, among other things, how there are a number of related but distinct questions to be answered in resolving these controversies.

One central question in these controversies is a question about which events mark the cessation-of-life (or the cessation-of-functioning-as-an-organism) of a human organism: are certain cardiopulmonary events the relevant events, or are certain neurological events the relevant events? This question is recognized by all sides to the controversies about when a human organism dies, though not all sides agree on how to best answer it. Furthermore, this question is directly relevant to the question of whether a given case is best described in terms of you "coming back from the dead": for example, if the relevant events are certain neurological events, then the fact that you "came back" after certain cardiopulmonary events occurred does not mean that you came back from the dead.

However, a second central question in these controversies is a question about the concept of death itself: does the concept of death include the notion of irreversibility, or does it make sense to say that someone died, momentarily, and then came back to life? Unlike the situation regarding the first question, this second question is sometimes *not* recognized by all sides to the controversies about when a human organism dies. Indeed (and again unlike the situation involving the first question), to the extent that different sides even recognize this second question to begin with, they seem to *agree* on how to best answer it: the answer they assume is that the concept of death *does* include the notion of irreversibility. Of course, this second question is even more relevant to the question of whether a given case is best described in terms of you "coming back from the dead": if the idea of coming back from the dead is incoherent, then *no* case should be described in those terms.

The two questions just mentioned are closely related. I might think that physical event P is the relevant event marking the death of a human organism, precisely

[2]The President's Council on Bioethics (2009).

because I think that this event marks the cessation-of-life of a human organism. But you might disagree, precisely because you think P is reversible.

One example of how these two questions are closely related is found in the President's Council's discussion of the organ donation protocols commonly referred to as "controlled donation after cardiac death" or "controlled DCD" protocols. As they describe the background,

> Human beings whose deaths have been determined according to the more traditional cardiopulmonary standard may also provide organs... .a potential nonheart-beating donor, in the vast majority of cases, is an individual who is ventilator-dependent but not yet deceased according to today's neurological standard. The ventilator is then removed, the patient is watched and kept comfortable until the heart stops circulating blood through the body, a waiting period is observed (usually two to five minutes), and then the surgical procurement of organs begins. . .[3]

The Council soon moves to a discussion of the question, "Are those who donate organs under a controlled DCD protocol actually dead at the time of donation?" They frame their answer this way:

> It might seem somewhat surprising that this is a matter of controversy. After all, at the time of procurement, the donor's heart has stopped and he or she is no longer breathing—either spontaneously or with ventilator support. Thus, the individual would seem to meet the first (more traditional) standard for determining death, that is, in the wording of the UDDA, "irreversible cessation of circulatory and respiratory functions."[4]

They then go on to explain how everything hangs all whether one understands the term "irreversible" in a strict sense or in a weaker sense:

> The difficulty here stems from the crucial requirement that cessation of circulatory and respiratory functions be irreversible. In truth, there is reason to doubt that the cessation of circulatory and respiratory functions is irreversible, in the strict sense, in every case of controlled DCD. To call the loss of functions irreversible, it must be the case that the functions could not possibly return, either on their own or with external help. It is often possible, however, to cause circulation and respiration to return by administering cardiopulmonary resuscitation (CPR). If this were attempted after the "declaration of death" in controlled DCD, some patients would indeed regain—for a brief time, at least—a heartbeat and some capacity to breathe. If this were to occur, the patient would certainly not have been "resurrected," but instead would have been (according to the cardiopulmonary standard of death) resuscitated, i.e., prevented from dying. Thus, the prior "declaration of death" would turn out to be questionable. The patient was, it could be argued, no more dead than a person who collapses in his or her home, loses heartbeat, and is resuscitated by paramedics who arrive moments later.[5]
>
> ... For this reason, many have argued that the word "irreversible" in this context should be understood in a weaker sense than that spelled out above: It should be understood to mean "cessation of circulatory and respiratory functions under conditions in which those

[3]The President's Council on Bioethics (2009, pp. 79–80).
[4]The President's Council on Bioethics (2009, p. 83).
[5]The President's Council on Bioethics (2009, pp. 83–84).

functions cannot return on their own and *will not* be restored by medical interventions." This looser sense of the term "irreversible" would seem to be a better fit in this context.[6]

So, then, if we take the term "irreversible" in a strict sense, then some donors in controlled DCD will not meet the traditional test for death ("irreversible cessation of circulatory and respiratory functions"), but if we take that term in a weaker sense, then perhaps all donors in controlled DCD will meet the traditional test. Notice, what stands behind this traditional test itself is the idea that death itself is, in some sense, irreversible, that the concept of death includes the concept of irreversibility.

Would switching to some neurological standard of death avoid this assumption about the irreversibility of death? I do not think that it would. Consider the neurological standard that the White Paper report ends up defending, which has been the dominant standard for the past several decades and which they call "total brain failure":

Although the choice of an appropriate term is important, it is more crucial to maintain a distinction between naming the medical diagnosis of a condition and declaring an individual dead on the basis of that medical diagnosis. In this report, we will employ the term "total brain failure" for the medical diagnosis. The precise meaning of "total" in this composite term is discussed in Chapter 3. Here, at the outset, we emphasize that total brain failure is, by definition, an irreversible condition. Thus, to be more explicit one could employ the term, "total and irreversible brain failure."[7]

So then, here as before, "irreversibility" is crucial. The reasons why it is crucial are spelled out in this suggestive passage:

...even the loss of all functions of the CNS [Central Nervous System] is not a sufficient criterion for declaring death if this loss of function is not irreversible. Again, there are critical care cases that demonstrate the importance of this qualification—for instance, when a patient is in a deep, non-breathing ("apneic") coma during a critical emergency and the support of the ventilator allows time for CNS functions to return. In some cases like this, a full recovery of CNS functions occurs. More often, though, the functions that return will only be enough to leave the patient in a "vegetative state" that, if it persists, will be labeled a PVS (a persistent vegetative state)...the point here is that the deep, non-breathing coma that the patient was in prior to "waking" into the vegetative state could not have been death since the loss of functions proved to be reversible.[8]

In other words, the reasons why irreversibility is crucial to "total brain failure" are reasons that refer both to empirical facts and to the concept of death itself.

A nearby passage in the report explains that a patient's *history* is also relevant to diagnosing "total" brain failure:

The cause of the patient's brain injury cannot be hypothermia, poisoning, drug intoxication, or any such cause that brings about metabolic changes that can mimic the effects of total brain failure. The reason that a total brain failure diagnosis is ruled out in these cases

---

[6]The President's Council on Bioethics (2009, p. 84).

[7]The President's Council on Bioethics (2009, p. 19).

[8]The President's Council on Bioethics (2009, pp. 29–30).

> is plain: A condition like this is often transient—it may clear up when the cause of the metabolic change passes out of the patient's system or is otherwise removed.[9]

In a footnote to this passage, the report cites a 2006 "clinical case study of a patient who showed all the signs of total brain failure after a snake bite but then recovered after receiving an antidote." This case study illustrates how the very same set of symptoms can be "the effects of total brain failure" in one case and yet can be "metabolic changes that can mimic the effects of total brain failure" in the next case. Therefore, the presence of these symptoms, by themselves, does not tell us whether the brain failure is total or transient.

The fact that a patient might show all the signs of total brain failure, and yet turn out to be such that his condition is reversible, and the fact that some of these reversible conditions might even require external assistance to reverse (e.g., the antidote in the snake bite case), are significant facts. What they show is that, even if we prefer a neurological standard of death, and even if we want to incorporate "irreversibility" in some sense into our diagnosis of death in a given case, we must be open to the possibility that future technologies might be able to reverse conditions that are currently considered irreversible. After all, there may have been a time when people thought that the condition caused by certain snake bites was irreversible. But the development of an antidote for these kinds of snake bites proved that the conditions were not entirely irreversible after all.

How do these discussions relate to the argument producing our "corpse problem"? A tempting strategy for trying to block this sort of argument is to claim that, *in any case* where there was a genuine revival or resuscitation, the individual was never dead to begin with. However, this "not dead yet" strategy is not the strategy I will take. In order for the "corpse problem" to even get off the ground, we must assume that the concept of death does *not* include the concept of irreversibility. In other words, we must assume that there is nothing incoherent about the concept of "reversible death", that there is nothing redundant about the concept of "irreversible death", and that there is nothing inconsistent about the idea of being "dead" at one time and "alive" at a later time. I am willing to accept these assumptions. I realize that this is swimming against the stream. But I think that the "not dead yet" strategy is not the right way out of the corpse problem. I recognize the need to respond to the corpse problem in a different way.

What about the question of which standard (cardiopulmonary or neurological) is the appropriate standard for determining when a living thing is no longer alive? My approach does not assume (or at least, I do not intend it to assume) a particular answer to this question. Perhaps my approach will be useful for arriving at an answer to this question. But it does not, as far as I can tell, assume such an answer at the beginning.

So, then, there were two questions surfaced at the outset of this discussion of the corpse problem. First, which events (cardiopulmonary or neurological) mark the cessation-of-life of a human organism? Second, does the concept of death include

---

[9]The President's Council on Bioethics (2009, pp. 31).

the notion of irreversibility? My approach assumes no answer to the first question but is willing to assume a negative answer to the second question.

3 Solving the Corpse Problem

The corpse problem can be solved. Note that the problem is generated by the assumption that the corpse itself is the locus, or bearer, of the relevant capacities—for example, the higher-order capacity to think. Since the locus or bearer of the relevant capacities is the locus or bearer of the serious moral status, it immediately follows that the corpse is the locus or bearer of the serious moral status.

But is it really true that the corpse is the locus, or bearer, of the relevant capacities? On first blush, this question appears to have a simple answer: of course the corpse is the bearer of the relevant capacities, since there is nothing else that has, or could have, these capacities. In part, this simple answer seems to rely on the assumption that, if you die, you *become* the corpse, and if you come back to life, then the corpse would *become* alive again. A parallel: if Reagan deteriorates in the ways mentioned in the hypothetical example, he *becomes* the brain-damaged human organism, and if Reagan recovers, then the brain damaged human organism *becomes* sentient and rational again.

The idea that you *become* the corpse at the moment of death may be implicit in the following statement from the White Paper Report: "Death is the transition from being a living, mortal organism to being something that, though dead, retains a physical continuity with the once-living organism."[10] This quote, on its surface, seems to assume that *you* survive your death as a cadaver or a corpse. However, another possible and more cautious reading of this quote is to view it as focusing on the sort of transition *your body* goes through, rather than the sort of transition *you* go through.

What I have called "the simple answer"—the idea that the corpse is the bearer of the relevant capacities, since there is nothing else that has, or could have, these capacities—is not obviously true. Indeed, I believe the simple answer is mistaken. We have reasons for believing that there is something else, besides the corpse, which has the relevant capacities. The reasons I shall focus on are those which emerge from a careful consideration of what are sometimes called "Near Death Experiences" or NDEs.

I recognize that someone who introduces NDEs as a topic for serious philosophical discussion risks being met with the kind of "incredulous stares" that philosopher David Lewis was notorious for getting (although he never got them for beliefs about NDEs). Still, I believe a serious consideration of this phenomenon has real philosophical value in this context.

There are several different kinds of Near-Death Experiences. They vary, for one thing, in their *reliability*—that is, whether we should trust what they purportedly

[10]The President's Council on Bioethics (2009, p. 17).

were experiences *of*. They also vary in their *relevance* to our present question—that is, the question of whether something besides the corpse might have certain capacities.[11]

Our consideration of the reliability and relevance of Near-Death Experiences will be enhanced by considering the first-hand reports of two very different men who had such experiences: Richard John Neuhaus and A. J. Ayer. One of the things that make these reports interesting is that both men were thoughtful, analytic, clear-thinking authors both before and after their Near-Death Experiences, and both were initially disposed not to put much stock in such experiences.

Richard John Neuhaus, before passing away in January 2009, was the President of the Institute for Religion and Public Life in New York, author of books such as *The Naked Public Square: Religion and Democracy in America*, and editor-in-chief of *First Things: A Monthly Journal of Religion and Public Life*.[12] Neuhaus had a sort of near-death experience in the intensive care unit of a hospital after his emergency surgery. He wrote about it in the introduction to a book he edited titled *The Eternal Pity: Reflections On Dying*. The section I quote is long, but I believe it is worth the effort:

> It was a couple of days after leaving intensive care, and it was night. I could hear patients in adjoining rooms moaning and mumbling and occasionally calling out; the surrounding medical machines were pumping and sucking and bleeping as usual. Then, all of a sudden, I was jerked into an utterly lucid state of awareness. I was sitting up in the bed staring intently into the darkness, although in fact I knew my body was lying flat. What I was staring at was a color like blue and purple, and vaguely in the form of hanging drapery. By the drapery were two "presences." I saw them and yet did not see them, and I cannot explain that. But they were there, and I knew that I was not tied to the bed. I was able and prepared to get up and go somewhere. And then the presences—one or both of them, I do not know—spoke. This I heard clearly. Not in an ordinary way, for I cannot remember anything about the voice. But the message was beyond mistaking: "Everything is ready now."
>
> That was it. They waited for a while, maybe for a minute. Whether they were waiting for a response or just waiting to see whether I had received the message, I don't know. "Everything is ready now." It was not in the form of a command, nor was it an invitation to do anything. They were just letting me know. Then they were gone, and I was again flat on

---

[11]First, however, consider an initial objection. It is sometimes claimed that Near-Death Experiences necessarily involve experiences of a person who has not died—otherwise these experience would not be called *Near*-Death experiences. I believe this claim can be resisted by reflecting on how death itself is regularly thought of as being like a boundary. One can be near to a boundary by being close to it on either side of it, or even on the boundary itself. For example, if I am traveling by car, and I call my wife on a cell phone to tell her that I am "near" the Indiana-Ohio border, this by itself does not tell her which side of the border I am on. Or consider the boundary of the California shoreline where the land meets the water. One can be "near" this boundary whether one is on the land or in the water (or both). A ship captain telling his crew not to steer "too near" the land, and a father telling his toddler not to walk "too near" the water, are both speaking of a state of being "near" the shoreline—just from different directions. Near death experiences, then, should be thought of as those experiences had by persons close to the boundary of death, on either side of that boundary (if there is more than one side), or even at the moment of death. If one has had a near death experience, this still leaves open the question of whether one actually died.

[12]Neuhaus (1984).

my back with my mind racing wildly. I had an iron resolve to determine right then and there what had happened. Had I been dreaming? In no way. I was then and was now as lucid and wide awake as I had ever been in my life.

Tell me that I was dreaming and you might as well tell me that I am dreaming that I wrote the sentence before this one. Testing my awareness, I pinched myself hard, and ran through the multiplication tables, and recalled the birth dates of my seven brothers and sisters, and my wits were vibrantly about me. The whole thing had lasted three or four minutes, maybe less. I resolved at that moment that I would never, never let anything dissuade me from the reality of what had happened. Knowing myself, I expected I would later be inclined to doubt it. It was an experience as real, as powerfully confirmed by the senses, as anything I have ever known. That was almost seven years ago. Since then I have not had a moment in which I was seriously tempted to think it did not happen. It happened—as surely, as simply, as undeniably as it happened that I tied my shoelaces this morning. I could as well deny the one as deny the other, and were I to deny either I would surely be mad.[13]

As these paragraphs illustrate, Neuhaus is a critical thinker about his own experience. He had a healthy dose of skepticism about near-death experiences in general. Then he had one of his own. What else could he do but attempt to make sense of it?

The way Neuhaus sought to make sense of his experience is similar to the way most of us seek to make sense of our experiences: by drawing upon his background beliefs about the world and his place in it. For Neuhaus, this meant drawing upon his Christian (Roman Catholic) beliefs; he argues that his messengers were angels preparing to take him to purgatory in order to undergo a period of preparation in order to meet God.

This interpretive framework, of course, does not subtract from the reality of the experience. *However* his experience is to be interpreted, *it* is something that really happened to him as an experience. Otherwise there would be nothing to interpret.

These sorts of experiences are not limited to people with a religious belief system. For example, the atheist philosopher A. J. Ayer, who taught for many years at Oxford, had a near-death experience towards the end of his life. He explained and tried to interpret this experience in an article titled "What I Saw When I Was Dead," in the August 28, 1988 issue of the *Sunday Telegraph*. He later wrote a follow-up article, "Postscript to a Postmortem," in the October 15, 1988 issue of *The Spectator*. Both articles are reprinted, under the title "My Death," in an anthology of articles about death written by philosophers.[14]

According to Ayer's attendants at the hospital, his heart was stopped for four minutes after an episode of choking, due to some smoked salmon going down the wrong way. He tells of three unusual experiences that occurred to him during the scramble to save his life. The first two experiences he does not actually remember, but he came to know about them from the testimony of other people who were

[13]Neuhaus (2000, pp. 31–32).

[14]The anthology, edited by John Donnelly, is *Language, Metaphysics, and Death* (2nd edition; New York: Fordham University Press, 1994), and Ayer's article is on pp. 226–236. All page references which follow refer to this reprinting of Ayer's article. Note, the first of the two pieces, in addition to being printed in *The Spectator*, was also published in the *National Review*. See Ayer (1988, pp. 38–40).

around him at the time. The third experience he does remember, and did not come to know about it from the testimony of others.

> The only memory that I have of an experience closely encompassing my death, is very vivid. I was confronted by a red light, exceedingly bright, and also very painful even when I turned away from it. I was aware that this light was responsible for the government of the universe. Among its ministers were two creatures who had been put in charge of space. These ministers periodically inspected space and had recently carried out such an inspection. They had, however, failed to do their work properly, with the result that space, like a badly fitted jigsaw puzzle, was slightly out of joint.

This vivid experience of Ayer's gets even more interesting when he switches from passive observer to active participant:

> I felt that it was up to me to put things right. I also had the motive of finding a way to extinguish the painful light. I assumed that it was signaling that space was awry and that it would switch itself off when order was restored. Unfortunately, I had no idea where the guardians of space had gone and feared that even if I found them I should not be allowed to communicate with them. It then occurred to me that. . .it had become customary, since the vindication of Einstein's general theory of relativity, to treat space-time as a single whole. Accordingly I thought that I could cure space by operating upon time.
>
> I was vaguely aware that the ministers who had been given charge of time were in my neighbourhood and I proceeded to hail them. I was again frustrated. Either they did not hear me, or they chose to ignore me, or they did not understand me. I then hit upon the expedient of walking up and down, waving my watch, in the hope of drawing their attention not to my watch itself but to the time which it measured. This elicited no response. I became more and more desperate, until the experience suddenly came to an end.[15]

Ayer's description of this third experience is detailed, interesting, and somewhat funny in places. Here is how he begins his interpretation of it:

> This experience could well have been delusive. A slight indication that it might have been veridical has been supplied by my French friend, or rather by her mother, who also underwent a heart arrest many years ago. When her daughter asked her what it had been like, she replied that all that she remembered was that she must stay close to the red light.[16]

Ayer thinks that the experience of his French friend's mother indicates that his own experience might have been "veridical," which seems to mean "not delusive" in this context. Perhaps the idea is that the fact that she had a similar experience is at least one piece of evidence in favor of the idea that his experience was not just his own private delusion.

Nevertheless, what Ayer eventually concludes is that his third experience, and indeed all three experiences, were indeed just his own private delusions. Here is how he puts it:

> On the face of it, these experiences, on the assumption that the last one is veridical, are rather strong evidence that death does not put an end to consciousness. Does it follow that there is a future life? Not necessarily. The trouble is that there are different criteria for being dead, which are indeed logically compatible, but may not always be satisfied together.

---

[15] Ayer (1994, pp. 228–229).

[16] Ayer (1994, p. 229).

In this instance, I am given to understand that the arrest of the heart does not entail, either logically or causally, the arrest of the brain. In view of the very strong evidence in favour of the dependence of thoughts upon the brain, the most probable hypothesis is that my brain continued to function although my heart had stopped.[17]

These experiences of Neuhaus and Ayer are, like many NDEs, rather interesting. They are also, like many NDEs, susceptible of various interpretations, some of which are entirely "natural" or material. Perhaps, as Ayer thinks, his brain was playing tricks on him. Perhaps, against what Neuhaus thinks, he was only dreaming.

A 2002 article in the journal *Resuscitation* by Sam Parnia and Peter Fenwick discusses the way different types of NDE during a cardiac arrest might be explained naturalistically, and finds several such explanations inadequate[18]:

The occurrence of lucid, well structured thought processes together with reasoning, attention and memory recall of specific events during a cardiac arrest (NDE) raise a number of interesting and perplexing questions regarding how such experiences could arise. These experiences appear to be occurring at a time when cerebral function can be described at best as severely impaired, and at worst absent. Although, under other clinical circumstances in which the brain is still functioning, it may be possible to argue that the experiences may arise as a hallucination in response to various chemical changes in the brain, this becomes far more difficult during a cardiac arrest. NDE in cardiac arrest appear different to hallucinations arising from metabolic or physiological alterations, in that they appear to occur in a non-functioning cortex, whereas hallucinations occur in a functioning cortex. Therefore, it is difficult to apply the same arguments for their occurrence. In addition cerebral localisation studies have indicated that thought processes are mediated through a number of different cortical areas, rather than single areas of the brain. Therefore a globally disordered brain would not be expected to produce lucid thought processes.[19]

At the end of their article, Parnia and Fenwick consider whether

. . .the occurrence of NDEs during a cardiac arrest, when the mind (the collection of all our thoughts, feelings and emotions) and consciousness (self awareness) appear to continue at a time when the brain is non-functional and clinical criteria of death have been reached, can be proven objectively through large studies. . .[20]

Interestingly, their answer is that "appropriate" studies of this sort may indeed be possible:

Such studies are currently possible, and it has been proposed to test the claims of 'consciousness' and being able to 'see' during cardiac arrest objectively by use of hidden targets that are only visible from a vantage point above. Although, at first these suggestions may sound rather unconventional, the study of consciousness has itself for many years been thought of as unconventional, but has now become a significant point of debate in neuroscience. Therefore, a new way of thinking may be needed to provide an insight into understanding this intriguing, yet largely undiscovered area of science.[21]

---

[17] Ayer (1994, p. 229).
[18] Parnia and Fenwick (2002).
[19] Parnia and Fenwick (2002, p. 8).
[20] Parnia and Fenwick (2002, p. 10).
[21] Parnia and Fenwick (2002, p. 10).

As the quoted passages from Parnia and Fenwick suggest, one sub-set of NDEs that are less easily explained by brain tricks or dreams are sometimes called Out-of-Body Experiences or OBEs. These conscious episodes are characterized by the experience of being outside of one's own body: perhaps by viewing things from a vantage point above the body, or even by being aware of things far away from the body. Since these experiences are not merely first-person reports of (say) seeing a bright light "responsible for the government of the universe" (to take Ayer's example), they are reports of things that other people are sometimes able to corroborate.

For example, Dutch researcher Pim van Lommel and colleagues, in a prospective multi-center study of NDE published in *The Lancet* in 2001, describe how in the pilot phase of their research in one hospital, a nurse who worked in a coronary-care unit reported an OBE of a resuscitated patient that was "veridical" even though the patient was in a deep coma during CPR (recall this was also the word Ayer chose to contrast with "delusive"). The story is worth quoting in full:

> During a night shift an ambulance brings in a 44-year-old cyanotic, comatose man into the coronary care unit. He had been found about an hour before in a meadow by passers-by. After admission, he receives artificial respiration without intubation, while heart massage and defibrillation are also applied. When we want to intubate the patient, he turns out to have dentures in his mouth. I remove these upper dentures and put them onto the 'crash car'. Meanwhile, we continue extensive CPR. After about an hour and a half the patient has sufficient heart rhythm and blood pressure, but he is still ventilated and intubated, and he is still comatose. He is transferred to the intensive care unit to continue the necessary artificial respiration. Only after more than a week do I meet again with the patient, who is by now back on the cardiac ward. I distribute his medication. The moment he sees me he says: 'Oh, that nurse knows where my dentures are'. I am very surprised. Then he elucidates: 'Yes, you were there when I was brought into hospital and you took my dentures out of my mouth and put them onto that car, it had all these bottles on it and there was this sliding drawer underneath and there you put my teeth.' I was especially amazed because I remembered this happening while the man was in deep coma and in the process of CPR. When I asked further, it appeared the man had seen himself lying in bed, that he had perceived from above how nurses and doctors had been busy with CPR. He was also able to describe correctly and in detail the small room in which he had been resuscitated as well as the appearance of those present like myself. At the time that he observed the situation he had been very much afraid that we would stop CPR and that he would die. And it is true that we had been very negative about the patient's prognosis due to his very poor medical condition when admitted. The patient tells me that he desperately and unsuccessfully tried to make it clear to us that he was still alive and that we should continue CPR. He is deeply impressed by his experience and says he is no longer afraid of death. 4 weeks later he left hospital as a healthy man.[22]

In summarizing the results of their prospective study, van Lommel and colleagues claim that a number of proposed theories for explaining NDE were inadequate to explain their data:

> Our results show that medical factors cannot account for occurrence of NDE; although all patients had been clinically dead, most did not have NDE. Furthermore, seriousness of the

[22] Van Lommel et al. (2001, p. 2041).

> crisis was not related to occurrence or depth of the experience. If purely physiological factors resulting from cerebral anoxia caused NDE, most of our patients should have had this experience. Patients' medication was also unrelated to frequency of NDE. Psychological factors are unlikely to be important as fear was not associated with NDE.[23]

They also refer to a case described by the American cardiologist Michael Sabom:

> Sabom mentions a young American woman who had complications during brain surgery for a cerebral aneurysm. The EEG of her cortex and brainstem had become totally flat. After the operation, which was eventually successful, this patient proved to have had a very deep NDE, including an out-of-body experience, with subsequently verified observations during the period of the flat EEG.[24]

Finally, they briefly consider the philosophical implications of their research:

> With lack of evidence for any other theories for NDE, the thus far assumed, but never proven, concept that consciousness and memories are localised in the brain should be discussed. How could a clear consciousness outside one's body be experienced at the moment that the brain no longer functions during a period of clinical death with flat EEG? Also, in cardiac arrest the EEG usually becomes flat in most cases within about 10 s from onset of syncope. Furthermore, blind people have described veridical perception during out-of-body experiences at the time of this experience. NDE pushes at the limits of medical ideas about the range of human consciousness and the mind-brain relation.[25]

My point in citing these anecdotes and studies is not to claim that NDEs prove that human beings have an immortal soul that can survive the death of their body. Nor is it to give the impression that the academic community has accepted such a claim. The debates about NDEs continue to this day, and even those who believe in them are divided on how to best interpret them and what their philosophical implications are. Even the *Lancet* article did not convince everyone, as the correspondence to that article indicates.[26]

On the other hand, my discussion of NDEs and OBEs does have several important implications for our discussion of capacities and moral status, both when considering temporary change scenarios involving death ("the corpse problem") and more broadly.

---

[23] Van Lommel et al. (2001, p. 2043).

[24] Van Lommel et al. (2001, p. 2044).

[25] Van Lommel et al. (2001, p. 2044).

[26] Some of the correspondence was serious, and some was whimsical. For example, the following humorous letter from Doctor Richard Couper echoes A. J. Ayer's response to his first Near-Death Experience:

> Sir—Pim van Lommel and colleagues' study reminds me of an apocryphal comment attributed to Kerry Packer, Australia's wealthiest man. Packer had a myocardial infarction while riding a polo pony. A nearby ambulance crew resuscitated him. Packer reported his experience with the telling comment: "Mate, I tell you there is nothing there". He was obviously not keen to repeat the experience and promptly equipped the New South Wales ambulance service with defibrillators. . .It is a pity that Kerry Packer, who, in his rare public utterances tells it as he sees it, could offer no further insight into the presence of the human soul" (Couper, 2002, p. 2116).

First, consider the temporary change scenarios involving human death. However "death" is defined, NDEs give us some evidence for the persistence of a self in between the time of death and the time of revival (this assumes, of course, that the notion of death does not contain the notion of irreversibility). Therefore, temporary annihilation of the self is ruled out *in those cases*: the self is still there, still endowed with important psychological capacities, and indeed, still exercising some of those capacities. Without temporary annihilation, there is no need to posit "temporal gaps" in the self during the time of death. However, the persistence of the self, by itself, does not escape the corpse problem: perhaps the so-called corpse is the entity having the experiences, and thus is the entity endowed with the capacity for experiences. So, if the so-called corpse itself has these capacities, and if capacities are the basis of serious moral status, then the so-called corpse still has serious moral status.

This is where a second implication becomes relevant: in the temporary change scenarios involving human death, however "death" is defined, those NDEs known as OBEs give us some evidence for the persistence of a *disembodied self with various psychological capacities* in between the time of death and the time of revival (once again, this assumes that the notion of death does not contain the notion of irreversibility). Therefore, in addition to ruling out temporary annihilation of the self, these cases also rule out embodiment during the interval of time between death and revival. Without embodiment, there is no need to posit survival-as-a-corpse in these cases. In other words, the thing with the relevant capacities during the period of time between death and revival is not the corpse, but the disembodied self. The locus, or bearer, of the capacities, is not the corpse, but the disembodied self. Therefore, since the bearer of the relevant capacities is the bearer of the serious moral status, it follows that the thing with the serious moral status during the time between death and revival is not the corpse, but the disembodied self.[27]

These two implications in the cases of the corpse problem are relevant to our broader question of how to approach apparent temporary change scenarios in various settings. The state of the discussion seems to be as follows. In all the possible apparent temporary change scenarios which we can imagine you going through, there are two main options from which to choose. Either (1) you persist through the apparent temporary change, or (2) you *do not* persist through the apparent temporary change.

Consider the first main option. If you persist through the apparent temporary change, either (1a) you persist as an object accessible to the investigative efforts of third-person observers, or (1b) you persist, *but not* as an object accessible to the investigative efforts of third-person observers.

Consider the second main option. If you do not persist through the apparent temporary change, either (2a) you cease to exist entirely—you might think this if you believe that you cannot have temporal gaps—or (2b) you *do not* cease to

---

[27] In this and what follows, I shall try to speak in terms of a (disembodied) "self" rather than a (disembodied) "soul", since I wish to avoid an intramural debate about whether you *are* your soul: even if you think that you are not your soul, you can agree that you, yourself (i.e., your "self"), can survive in a disembodied state.

exist entirely—you might think this if you believe that you can have temporal gaps.

I have assumed that you cannot have temporal gaps. That means the only options to pick from, in any apparent temporary change, are options (1a), (1b), and (2a). In some cases of apparent temporary changes, it is very difficult to empirically settle the question between these options. However, in one of the cases where we can gather some empirical evidence to help us settle the question—namely, the case of those who have apparently died and come back to tell about it—we have some evidence that a person does persist through the apparent temporary change. In some of *these* cases—namely, those cases of those who have apparently died and come back to tell about it, and where we have some evidence that a person does persist through the apparent temporary change—the evidence indicates that the person persists, *but not* as an object accessible to the investigative efforts of third-person observers. In other words, these cases support option (1b).

In the apparent temporary change scenarios involving human death, we have evidence for the persistence of a self in between the time of death and the time of revival. Therefore, annihilation—option (2b)—is effectively ruled out in those cases. (Or, to speak more guardedly, annihilation should not be assumed unless there are compelling reasons for doing so. One can always bite the bullet, I suppose, and say that the original self was annihilated at the moment of death, and that a new self, with beliefs, memories, character traits, and so on, that perfectly match those of the original self, comes into being at a time between the moment of death and the moment of resuscitation.)

Likewise, in the apparent temporary change scenarios involving human death, we have evidence for the persistence, not just of *a* self, but of a *disembodied* self in between the time of death and the time of resuscitation. Therefore, survival-as-a-corpse—option (1a)—is effectively ruled out in those cases. (Or, again, to speak more guardedly, survival-as-a-corpse should not be assumed unless there are compelling reasons for doing so.)

Therefore, the thing with the relevant capacities—the capacity to think, the capacity to hear, the capacity to see (van Lommel even mentions a "blind" person getting back the faculty of sight during the OBE), the capacity to hear, to remember, to love, and so on—during the time between death and revival is not the corpse, but the disembodied self.

Therefore, the thing with the serious moral status during the time between death and revival is not the corpse, but the disembodied self.

In his own personal statement towards the end of the White Paper Report, Edmund Pellegrino makes the passing remarks that

> ...the metaphysical definition of death as separation of the body from its vital principle is still held as the authoritative definition by many worldwide. Plato put it most bluntly: "Death in my opinion is nothing else but the separation from each other of two things, soul and body." No precise congruence of this concept with any observable set of clinical facts has ever been agreed upon.[28]

---

[28]The President's Council on Bioethics (2009, p. 110).

One of the things that Pellegrino's remarks suggest is this: we do not have enough empirical evidence, at present, to know precisely at what point the soul, or self, separates from the body. Put differently and in the language of capacities, we do not have enough empirical evidence, at present, to know precisely at what point the locus of capacities shifts from being embodied to being disembodied (or even if there is such a point that is the same point for all human beings).

However, this deficit of evidence is not, in principle, irresolvable. The structure of the evidence-gathering can be straightforwardly described. If we were to do experiments in which we gradually brought a human adult through changes in bodily functioning, and we discovered that patients characteristically floated out of their bodies when a certain step in the process was taken, and then floated back into their bodies when that step was reversed, then we would have evidence for that step being the point in the process in which the human self leaves the body. Of course, such experiments sound rather gruesome, like something out of the 1990 movie *Flatliners*, in which medical students try to induce Near-Death Experiences in one another. We do not experiment with human lives this way, even though we might learn a good deal about the nature of human selves if we did so. On the other hand, the prospective studies cited above illustrate that scientific research into these things can go on without violating medical ethics. If a researcher tapes a hidden message on the top of a shelf in the coronary care unit of a hospital, and then simply keeps records of whether certain patients with certain physiological conditions report "seeing" the message while "floating" out of their bodies, this would not obviously violate any principles of biomedical ethics.

We do not have any empirical evidence, at present, of how to handle the thousands of temporary change scenarios that we might imagine. But the structure of the argument in these cases would apparently be the same as the structure of the argument in cases involving temporary death. For example, if we were to do experiments in which we gradually replaced the tissues of a human adult with the tissues of a non-human organism, and we discovered that patients characteristically floated out of their bodies when a certain step in the process was taken, and then floated back into their bodies when that step was reversed, then we would have evidence for that step being the point in the process in which the human self leaves the body. Of course, such experiments are against the law, and for good reasons. We do not tolerate scientific experiments like this, even though we might learn a good deal about the nature of human selves if we did so.

One of the upshots, however, for the present inquiry is this: it is not just metaphysical speculation that informs our judgments about what would happen to us in various "temporary change" scenarios. Rather, it is our experiences of "going through" various temporary change scenarios ourselves, and our experiences of hearing the stories of others who have "gone through" various temporary change scenarios themselves, that informs our overall picture of what might happen, and what must happen, in a given temporary change scenario. I make no pretense to be an expert on what will happen if we ever decide to conduct such experiments. When we begin to investigate the details of such near-death-experiences—or in the case of turning us into a non-human animal, what might be called near-animalization-experiences (or re-speciation experiences?)—the complexities and challenges they pose to our theories of human identity may be deeply perplexing. But my point is that we should listen to what our best empirical evidence tells us when formulating our views about capacities, moral status, and the relationship between the two.

As I stated above when introducing the corpse problem, my own preferred method, when approaching apparent temporary changes, is to aim for an optimal reflective equilibrium between our intuitions, our metaphysical theories, and our experiences (both first- and third-person). There is some unavoidable give-and-take in trying to achieve such equilibrium: accepting certain intuitions or experiences requires rejecting or adjusting a given theory, and accepting a given theory requires rejecting certain intuitions or experiences. My attempt to use Near-Death Experiences to avoid the corpse problem illustrates how this can be done.

4 Are Active Capacities Preferable to Passive?

Another common objection to the view defended in these pages is that the view does not do enough to distinguish organisms from other sorts of beings because it relies on "passive" as well as "active" higher-order capacities. In contrast, the view of human capacities and their moral relevance defended by contemporary "natural law" philosophers (like Robert P. George, John Finnis, and Germain Grisez) relies only on "active" higher-order capacities and avoids "passive" higher-order capacities. As a result of their emphasis upon "active" capacities, these thinkers are able to distinguish between organisms and other sorts of beings. For example, organisms, via certain "active" capacities, can be said to be "executives" of their own development (this phrase is originally from William Hurlbut, a member of the President's Council on Bioethics), whereas other sorts of beings who lack these active capacities cannot be said to be "executives" of their own development.

In this section, I would like to respond to this concern. I shall argue that my view, while very similar to the contemporary natural law view, actually does a slightly better job than it when considering certain problem cases. Thus my view should be seen as a friendly amendment to the natural law view, which allows it to fruitfully address what might otherwise be a number of counterexamples.

A recent and representative example of the contemporary natural law view is found in Robert P. George and Christopher Tollefsen's book *Embryo: A Defense of Human Life*.[29] In a key passage on the nature of the human embryo, they emphasize three of its features: its distinctness, its humanness, and its wholeness. They rely on the concept of an "active disposition" when explaining the third of these features:

> ...the human embryo is a complete or whole organism, though immature. The human embryo, from conception onward, is fully programmed and has the *active disposition* to use that information to develop himself or herself to the mature stage of a human being,

[29] George and Tollefsen (2008).

> and, unless prevented by disease or violence, will actually do so, despite possibly significant variation in environment (in the mother's womb). None of the changes that occur to the embryo after fertilization, for as long as he or she survives, generates a new direction of growth. . .Rather, all of the changes (for example those involving nutrition and environment) either facilitate or retard the *internally directed growth* of this determinate and enduring individual.[30]

Part of what supports these claims about the "wholeness" of the human embryo is a commonsense idea about persistence through time: namely, the idea that the human embryo is the same individual as the fetus, child, adolescent, and adult that it will (hopefully) one day develop into:

> The embryonic, fetal, child, and adolescent stages *are just that*—stages in the development of a determinate and enduring entity—a human being—who comes into existence as a single-celled organism (a zygote) and develops, if all goes well, into adulthood many years later.[31]

Another part of what supports these claims about "wholeness" is the idea that the human embryo is endowed with two crucial resources—namely, genetic material/information and an active disposition to use it—that give the embryo an ability to transform itself into the fetus, child, adolescent, and adult it will (hopefully) one day develop into:

> Human embryos possess the genetic and epigenetic primordia for self-directed growth into adulthood with their unity, determinateness, and identity fully intact. . .the human embryo possesses all of the genetic material needed to inform and organize its growth. The direction of its growth is not extrinsically determined, but is in accord with the genetic information within it. Moreover, unless deprived of a suitable environment or prevented by accident or disease, the embryo is *actively developing itself* to maturity. Thus, it not only possesses all of the necessary organizational information for maturation, but it truly possesses an *active disposition* to develop itself using that information.[32]

Although these passages are not explicit on the matter, they do suggest that an entity possessing merely the organizational information for maturation (the genetic and epigenetic primordia) but not the active disposition to develop itself according to such information would thereby lack one of the two crucial resources of a human embryo, and thus would not be a human embryo. And indeed, George and Tollefsen recognize that their emphasis upon an active disposition is not only useful for *identifying* the human embryo with the human adult it will one day develop into. In addition, it is also useful for *distinguishing* the human embryo from the human

---

[30]George and Tollefsen (2008, p. 50), emphasis mine. The same passage is found, almost verbatim, in Lee and George's (2004) article. The only interesting difference for our purposes is that the 2004 article uses the phrase "The human embryo. . .is fully programmed *actively to develop* himself or herself. . ." (p. 14), while the 2008 book uses the phrase "The human embryo. . .is fully programmed *and has the active disposition to use that information to develop* himself or herself. . ." (emphasis mine in both cases).

[31]George and Tollefsen (2008, pp. 50–51).

[32]George and Tollefsen (2008, pp. 52, 53), emphasis mine.

parts it was at one time produced from, and from other entities produced by such parts. Consider, for example, the following passage:

> If the embryo were not a complete organism, then what could it be? Unlike the spermatozoa and the oocytes, it is not merely a part of a larger organism, namely the mother or father. Nor is it a disordered growth, such as a hydatidiform mole or a teratoma. Such entities lack the internal resources *actively to develop themselves* to the next more mature stage of the life of a human being. Their direction of growth is, like a cancer, not toward maturity.[33]

In a related passage just a page latter, George and Tollefsen do not explicitly use the term "active" or "disposition," but they do emphasize how the embryonic organism's development is directed "from within" rather than from anything "extrinsic."[34] Consider also the following nearby passage, which links the idea of active dispositions with the idea of internal self-direction in order to distinguish embryos from somatic cells:

> Like a male or female gamete, a somatic cell is functionally only a part of a larger organism. It does not contain within itself the internal resources and *active disposition* possessed by the embryo to develop itself to its full maturity. The gene expression pattern of the nucleus of a somatic cell may be transformed by the process of Somatic Cell Nuclear Transfer, so as to produce a human embryo. But such a change must come from without; skin cells never, by internal self-direction, develop into human embryos. No skin cell possesses an *active disposition to use the genetic information within it to develop itself toward maturity* as a complete and separate member of the species.[35]

I would like to begin evaluating the position of George and Tollefsen. Perhaps the first thing to say is that there is much to recommend in these passages. George and Tollefsen are right to emphasize the active disposition of a human embryo to develop itself towards maturity. If we accept my stipulation above that a disposition is the same thing as a capacity,[36] then this emphasis of George and Tollefsen amounts to an

---

[33]George and Tollefsen (2008, p. 53), emphasis mine.

[34]George and Tollefsen (2008, p. 54).

[35]George and Tollefsen (2008, pp. 54–55), emphasis mine. The 2004 article by Lee and George has a parallel passage emphasizing active dispositions: "Nor are human embryos comparable to somatic cells (such as skin cells or muscle cells), though some have tried to argue that they are. Like sex cells, a somatic cell is functionally only a part of a larger organism. The human embryo, by contrast, possesses from the beginning the internal resources and active disposition to develop himself or herself to full maturity; all he or she needs is a suitable environment and nutrition. The direction of his or her growth *is not extrinsically determined*, but the embryo is internally directing his or her growth toward full maturity" (Lee and George, 2004).

[36]As indicated in Chapter 1, Tollefsen has indicated to me (in correspondence) that he would prefer to make a distinction between *dispositions*—which, on his proposal, are potentialities that can be possessed by all manner of things—and *capacities*—which, on his proposal seem to be dispositions of living things. However, as I indicated earlier, I am reluctant to adopt Tollefsen's proposal here. Both philosophical usage and common usage of the dispositional family of terms is highly variable. Therefore, I think it remains preferable to stick with my stipulation that I will be using all such dispositional terms as synonymous with "capacities", and that I will mark distinctions among capacities by adding extra terms like "active" and "passive". I believe that simply keeping this difference between Tollefsen and me in mind will go some way towards properly framing the extent of our debate over the issue of active and passive capacities.

emphasis upon the embryo's active capacity to develop itself towards maturity. Also, once we realize that maturity for a human being includes thousands of activities like thinking, walking and loving, we realize that the embryo's active capacity to develop itself towards maturity is, in many ways, a convenient shorthand for its possession of a higher-order capacity to think, a higher-order capacity to walk, a higher-order capacity to love, and so on. Finally, it is worth noting that the moral conclusion George and Tollefsen eventually work toward in Chapter 4 of their book is rather similar to the moral conclusion I have worked toward in Chapter 2 of the present book.[37] However, I believe that there are several reasons for supplementing George and Tollefsen's account with an explicit recognition of the metaphysical reality and moral importance of what I have called "passive" capacities.

To begin with, many passages from George and Tollefsen contain clauses beginning with the word "unless." For example: the embryo will develop itself to the mature stage of a human being "unless prevented by disease or violence" (50); an embryonic human being will develop itself to the fetal stage "[u]nless severely damaged or denied or deprived of a suitable environment" (50–51); the embryo is actively developing itself to maturity "unless deprived of a suitable environment or prevented by accident or disease" (53). These "unless"-clauses raise the following objection: if the possession of an active capacity/disposition of the relevant sort is *required* to make an entity a whole human organism (rather than a mere part of another organism, or a teratoma, etc.), then how are we to avoid the conclusion that the entities referred to by these "unless"-clauses—namely, diseased entities, severely damaged entities, entities denied or deprived of a suitable environment—are not whole human organisms? The natural way of avoiding this objection is to supplement George and Tollefsen's account with an explicit recognition of passive dispositions/capacities. The entities referred to in the "unless"-clauses presumably still have the passive disposition/capacity to be developed to the more mature stages of a human being. This supplementation would allow the entities referred to by these "unless"-clauses to still count as whole human organisms.

George and Tollefsen might plausibly reply here that even in the cases referred to by the "unless"-clauses, the entities still have the relevant active dispositions. True, the embryo will develop itself to the mature stage of a human being "unless prevented by disease or violence" (50): but the embryo still has the active disposition to so develop itself *even when* prevented by disease or violence. True, an embryonic human being will develop itself to the fetal stage "[u]nless severely damaged or denied or deprived of a suitable environment" (50–51): but this being has the active disposition to so develop itself *even when* severely damaged or denied or deprived. True, the embryo is actively developing itself to maturity "unless deprived of a suitable environment or prevented by accident or disease" (53): but the embryo still has the active disposition to develop itself toward maturity *even when* deprived of a suitable environment or prevented by accident or disease. In general, the embryos still

[37] George and Tollefsen (2008, p. 86).

have the active disposition to X (for example, to develop towards maturity), but in the cases referred to by the "unless"-clauses, the embryos will not in fact X because of some contingent misfortune. The active disposition or capacity is still there, but it is merely blocked from expressing itself.

This strikes me as a plausible reply. However, this move seems to interpret the idea of an active disposition fairly widely. Indeed, it seems to interpret an active disposition so widely that it ends up swallowing up the metaphysical space that has often been occupied by what I would call a passive disposition. I am not convinced that this move is necessary, and to see why, let me recall some of the earlier discussions of passive potentiality in this book. In the Introduction, I remarked that the active/passive distinction was "somewhat rough" and I suggested an illustration of this distinction: I have an active capacity to raise my arm on purpose, and a passive capacity to feel pain when pricked. In Chapter 1, I noted how Aristotle thought that the primary sorts of dispositions or capacities were "active" because their locus was "in the agent, e.g. heat and the art of building are present, one in that which can produce heat and the other in the man who can build".[38] (In fairness to Aristotle, his favored term here was "potencies", and he reserved "disposition" and "capacity" for more technical uses.) Corresponding to these active powers were "passive" ones, so-called because their locus was "in the thing acted on...[e.g.] that which is oily can be burnt, and that which yields in a particular way can be crushed."[39] Such potencies "of being acted on" were still genuine because they are "the originative source, in the very thing acted on, of its being passively changed by another thing or by itself *qua* other".[40] Building on Aristotle here, I claimed that "active" capacities are powers to change things, whether in the world or in the bearer of the capacity, while "passive" capacities are powers for being changed by things in the world or in the bearer of the capacity.

Especially given this Aristotelian background, I believe that in the embryonic cases referred to by the "unless"-clauses of George and Tollefsen, it is just as good (indeed, preferable) to think of them as having a "passive" disposition to be developed toward maturity as it is to think of them as having an "active" disposition to develop themselves toward maturity. To see why, recall the emphasis George and Tollefsen place upon the *internality* of capacities. This emphasis upon internality is one that should take into account passive capacities as well. After all, passive capacities are, in a fairly straightforward sense, internal to the thing which has them. As Aristotle put it, such passive powers are "*in* the thing acted on..." To see this, consider a non-biological example. Two different rectangular keys (notes, planks) on a musical instrument like a xylophone, when struck with the same mallet, make sounds of quite different pitches. Why? Surely the difference in pitch is not due to anything "external" to the keys. Rather, the difference in pitch is due to the passive capacities of the keys, which are also internal capacities of the keys. One key has

[38] Aristotle (1941, 1046a26-28).

[39] Aristotle (1941, 1046a22-26).

[40] Aristotle (1941, 1046a12-14).

the internal passive capacity to resonate at such-and-such pitch when struck with the mallet; the other key has the internal passive capacity to resonate at a different pitch when struck with the mallet. The passivity of these capacities does not threaten the internality of these capacities.

But more can be said. In non-embryonic cases, it is often just as good (indeed, preferable) to think of an entity as having a "passive" disposition of some sort (for example, the passive disposition to be developed toward maturity) as it is to think of that entity as having an "active" disposition (for example, to develop themselves toward maturity). In Chapter 2, I argued that even if you do not have a functioning cerebral cortex for a period of time due to serious but temporary upper brain damage, you nevertheless retain the capacity (at some order or other) to think during this period of time, and this capacity may be a "passive" higher-order capacity rather than an "active" one, if the damage to the cortex requires a good deal of external assistance, such as surgery, to fix. Other philosophers have also noticed that for a patient in a reversible coma to still count as possessing a capacity for thinking, it is necessary to understand this capacity as a passive potentiality rather than an active one. Now, if these cases involving human adults are best accounted for by recognizing a passive (rather than active) capacity of some sort in the organism, then why aren't the analogous cases involving human embryos also best accounted for by recognizing a passive (rather than active) capacity of some sort in the organism? My point here is that there seems to be a need to invoke passive capacities in order to explain how certain adult human organisms are still bona fide organisms. Once this point is in place, the same approach for human embryos in comparable situations appears appropriate.

Still, I recognize that passive capacities are often resisted for a cluster of related reasons. First, people think that active capacities are necessary and sufficient for doing certain philosophical work, like making an organism an organism. Second, people think that passive capacities are (at best) superfluous and impotent, and (at worst) downright liabilities which muddy the water and open up new problems. I have tried in earlier chapters to address the second concern. My main line has been that allowing passive potentialities to bear moral weight is unproblematic as long as the distinction between identity and material constitution is upheld. The passive potentiality *a* has to become *b* in an identity-preserving way, and the passive potentiality *a*'s matter has to constitute *b*, are different sorts of potentialities. A living human adult in a profound coma has the former sort; a living human skin cell has only the latter sort. Still, let me approach this question from a slightly different angle.

Up until this point in the book, I have distinguished between what I called "identity-preserving" capacities and what I called "compositional" capacities, and I have only applied the active-passive distinction to the "identity-preserving" capacities. This may have given the impression that the active-passive distinction simply does not apply, or perhaps cannot apply, to compositional capacities. But this impression would be misleading. Even if there is not a perfect active-passive distinction to be found among compositional capacities, there seems to be something like it that deserves attention.

Recall the example of how a lump of bronze has the compositional capacity to "become" a statue of Socrates. This compositional capacity is probably a passive capacity, or at least towards the passive end of the spectrum. Now consider a thought experiment. Imagine a planet which has a kind of stuff that looks and behaves like our bronze, with one major difference: when this stuff is blown out of volcanoes in a hot liquid state, its internal chemical composition leads it to naturally cool into the shape of statues of Socrates. It does this regularly, on its own, without any help from intelligent craftsmen. The mountainsides of this planet are littered with cooled shapes of this stuff that have the same form as statues of Socrates on our planet. Arguably, this stuff would have the compositional capacity to "become" a statue of Socrates. But in this case the compositional capacity is probably an *active* capacity, or at least towards the active end of the spectrum.

Consider a second example, this time of living stuff in a living organism. Imagine a type of organism that is genetically programmed to undergo binary fission exactly every 18 years, but not a day sooner or later. For as long as anyone can remember, this type of organism has "reproduced" its kind in this way. This should not lead us to say that there is no organism present up until the moment of division. Neither should it lead us to say that there must be two organisms present before the moment of division. It is perfectly intelligible to say that there is exactly one organism present before the moment of division, that there are exactly two organisms present after the moment of division, and that neither of the two organisms after the moment of division is identical to the organism before the moment of division. Now, what is the best way of characterizing the moment of division in terms of active capacities? Well, on the one hand, this organism seems to have the internal, active disposition to (among other things) divide in two on its 18th birthday, thereby "becoming" two new organisms. That is an interesting idea in its own right. On the other hand, what is more interesting for my present purposes is this: there is a way of focusing in on the *stuff of which the organism is composed* before the moment of division. The stuff of which the organism is composed arguably has the capacity to constitute two organisms. Put differently, the stuff on the left-hand side of the organism has the capacity to constitute a new organism, and the stuff on the right-hand side of the organism has the capacity to constitute a new organism. But consider: is the capacity of this stuff, to constitute a new organism, an active capacity or a passive capacity? We might be tempted to say "passive!" But wait: the biochemical signals that are churning away within this stuff, propelling it in the direction of its new configuration, seem to be everything we could ask for in an active capacity. The chemicals, let us stipulate, are not responding to external signals, or at least are not determined by such "extrinsic" signals. The stuff as a whole is chugging right along according to its own internal clock of when it is time to divide and constitute two (new) organisms.

The point of these thought experiments is to show that the fact that a capacity is an *active* capacity is not, by itself, enough to guarantee that it is an *identity-preserving* capacity. The capacity of the bronze-like stuff is an active capacity to constitute statues of Socrates—but this is not an identity-preserving capacity because the bronze is not literally numerically identical to the statue (constitution is not the same relation as identity). Likewise, the capacity of the organic stuff in the pre-fission organism is an active capacity to constitute two post-fission organisms, but this capacity is not an identity-preserving capacity because the organic stuff is not literally numerically identical to the post-fission organisms (constitution is not the same relation as identity).[41]

More generally, determining whether a capacity is an active or a passive capacity is not at all the same thing as determining whether a capacity is an identity-preserving or a compositional capacity. Some writers seem to assume that a capacity is compositional just in case (if and only if) it is passive. Some writers seem to assume that a capacity is identity-preserving just in case (if and only if) it is active. Some writers seem make both assumptions at the same time. But both of these assumptions are mistaken.

When we think about the extremes of capacities using the distinction between active and passive capacities and the distinction between compositional and identity-preserving capacities, we sometimes only think about two out of the four possible extremes. We think, first, about an entity whose capacities are entirely active and identity-preserving: perhaps like Aristotle's god, a being of pure activity. Or we think, secondly, about a kind of stuff whose capacities are entirely passive and compositional: perhaps like Aristotle's bare matter, a mixture of pure passivity.

However, there is nothing incoherent in the idea of an entity whose capacities are entirely passive but identity-preserving. This entity would have no active capacities whatsoever, its changes would be entirely at the beck and call of other entities, and all of its change, growth, maturation, and so on would be directed from without. But the entity itself would remain the self-same individual over time, and indeed would never go out of existence: although it could change, all its changes would be identity-preserving changes. Likewise, there is nothing incoherent in the idea of a kind of stuff whose capacities are entirely active but compositional. This stuff would be constantly changing, constantly constituting differing things, but would never itself persist in the way you and I persist. And yet the changes from one form to the next could be as determined from within the stuff as you want: maximally internal, minimally extrinsic.

We human beings, of course, are somewhere in between these four extremes. We have some active, identity-preserving capacities and some passive, identity-preserving capacities. We are made up out of stuff that has some passive, compositional capacities and some active, compositional capacities. It is partly for this reason that I am concerned to highlight both our active capacities and our passive capacities, and to not allow one to push out the other.

---

[41] If one prefers, the last sentence of this paragraph could focus on just half of the stuff in the pre-fission organism: the capacity of the organic stuff in the left-hand side of the pre-fission organism is an active capacity to constitute one post-fission organism, but this capacity is not an identity-preserving capacity because the organic stuff is not literally numerically identical to the post-fission organism (constitution is not the same relation as identity). And the same goes for the right-hand side of the pre-fission organism.

Before concluding this comparison of my view with the recent natural law views of those like George and Tollefsen, it is worth noting that my views are compatible with, and indeed not obviously different than, those of contemporary natural law philosopher Alfonso Gómez-Lobo. Gomez-Lobo is a professor of philosophy at Georgetown University, a member of the President's Council on Bioethics, and has views on human persons very close to the views of George and Tollefsen. In a recent paper titled "Sortals and Human Beginnings", Gomez-Lobo's argument includes two points that are relevant to the present discussion.

First, in discussing whether a living human organism could survive past the death of the "person," Gomez-Lobo makes a comparison between cornea tissue transplants and brain tissue transplants. This comparison seems to assume that an individual whose higher brain functions are destroyed is still a human organism, and is still a person:

> My own view (derived indirectly from a statement of the American Academy of Neurology) is that it is more plausible to think that after the destruction of the cortex the person remains alive, although in a deep coma or state of permanent unconsciousness. One way to make this slightly more palatable could be as follows: a person who loses the cornea of both eyes because of illness or accident will become blind. She will not be able to exercise the capacity to see due to a severe lesion in her organs of sight. Imagine that corneal epithelial cell transplants have become successful, that the person undergoes transplantation, and can now see again. One could interpret her condition as a case in which the organism has retained what could be called the "deeper-lying faculty of sight" although the organ required to actualize it has been severely damaged. If the organ is replaced or repaired, the faculty can be actualized again. . .Imagine now the same situation for the cortex or higher brain. If the neurons lost because of hypoxic-ischemic encephalopathy could be replaced by deriving them from non-embryonic stem cells taken from the patient herself, it is not inconceivable in principle that the person could regain consciousness. If the organ has failed but is restored to its original condition, why couldn't the higher function that relies on that organ be recovered? This, of course, is not an argument. It is just a thought experiment, but not of the science fiction sort that has become common these days. Indeed, there is evidence that cornea transplants are already being performed.[42]

I believe that Gomez Lobo's thought experiments here are right on target. But notice, the thought experiments involve temporary change scenarios in which a lower-order capacity for sight (or thought) is lost, and then regained, while a higher-order capacity for sight (or thought) is retained throughout. What Gomez-Lobo calls the "deeper-lying faculty of sight" seems to be precisely the sort of higher-order capacity that I have been discussing throughout this book. And notice, further, that this higher-order capacity is not really an *active* higher-order capacity, but a *passive* higher-order capacity. The thought experiment assumes that the higher-order capacity for sight is still present, even when it would take epithelial cell transplants for the lower-order capacity for sight to return. The thought experiment assumes that the higher-order capacity for thought is still present, even when it would take

[42]Gomez-Lobo (2009, p. 7).

regenerative medical therapies to replace the neurons in the cortex. These higher-order capacities—to see in one case and to think in the other—are most plausibly understood as *passive* higher-order capacities: intrinsic receptivities to identity-preserving alterations. Of course, this says nothing at all about whether the immediate capacity for sight is a passive capacity (I'm inclined to think that it is), or whether the immediate capacity for thought is a passive capacity (I'm inclined to think that it isn't).

Second, in a footnote towards the end of his paper, Gomez-Lobo makes the claim that a fused genome is not strictly necessary for the existence of a human organism:

> Strictly speaking the fusion of the pronuclei of the gametes does not take place immediately after fertilization. Cf. Campbell and Reece (2002) 1002: "In contrast to sea urchin fertilization, the haploid nuclei of sperm and egg do not fuse immediately in mammals. Instead, the envelopes of both nuclei disperse, and the chromosomes from the two gametes share a common spindle apparatus during the first mitotic division of the zygote. Thus, it is not until after the first division, as diploid nuclei form in the two daughter cells, that the chromosomes from the two parents come together in common nuclei to form the genome of the offspring." I don't see any major philosophical difficulty in accepting a time lag between the actual co-presence of the key genetic factors contributed by the gametes and their actual fusion. The one-cell zygote containing 46 chromosomes is already different from the gametes and is programmed to take the next step, i.e. the mitotic division.[43]

Gomez-Lobo is here admitting that a fused human genome is not strictly necessary for a human organism to exist. Since Gomez-Lobo argues that you are a human organism, the admission that a fused human genome is not strictly necessary for a human organism to exist has the implication that a fused human genome is not strictly necessary for you to exist. It is enough for Gomez-Lobo that the key genetic factors contributed by sperm and egg are co-present, even if they are not actually fused.

Consider what happens when these two points from Gomez-Lobo are combined. First, a higher-order capacity can still be present in an entity, even when it takes significant external assistance in the form of surgical interventions to repair the entity to the point at which the entity can exercise the capacity. Second, a human organism can exist even when and where a fused human genome does not. These two points, when combined, seem to lead to a conclusion that I mentioned earlier in the book: even an entity that lacks a complete human genome can still be a human organism and can still be endowed with the higher-order capacities that are typical of an adult human organism. This does not mean that all bets are off concerning what sort of thing can be a human organism. But it does mean that a simple genetic reductionism will be inadequate to capture the nature of a human person. I suggest that the difference between my view and the new natural law view is not so very great after all.

---

[43] Gomez-Lobo (2009), Footnote 41, pp. 21–22.

5 Drawing Lines Near Altered Nuclear Transfer and Anencephaly

One set of concerns that many people have expressed to me[44] is that my view risks turning certain *obviously* non-human entities into human organisms, and that my view too easily turns certain *debatably* non-human entities into human organisms. For examples of the first type, there are some entities in utero that are clearly not human beings—entities like teratomas and hydatidiform moles. For examples of the second type, the products of ANT-OAR (Altered Nuclear Transfer-Oocyte Assisted Reprogramming) are debatably non-human entities. They are possibly in the exact same boat as teratomas and hydatidiform moles, although it is more open to debate whether the products of ANT-OAR are more like human embryos.

Now, it is not so controversial to claim that, if adequate genetic modifications could be made to either the obviously non-human entities, or to the debatably non-human entities, perhaps these entities could "become" human beings in the same way that sperm and egg can "become" human beings. But what is controversial is the claim that such entities are *already* human beings. Does not my line of argument force me into precisely such a controversial claim? Does not my line of argument imply that these entities are already human beings, since my line of argument wants to make room for genetically handicapped human infants and adults to count as human beings? The general question implicit in these particular concerns can be stated this way: where is the line between "defective" human beings, and entities so deficient that they should not be considered to be human beings in the first place?

These strike me as worthy concerns. However, I believe that these concerns can be answered. The first point to make is that the teratoma and the hydatidiform mole are not human beings because they do not seem to be human according to the definition of "human" given in Chapter 1. These entities are not "human" because they are not "an organism that has possessed, at some time in its past history, a structure made up out of one or more cells that have the same basic genotype as your cells and my cells have right now." This of course pushes us back to the question of the exact meaning of the phrase "the same basic genotype." The provisional definition I gave in Chapter 1 was that you and I have the same basic genotype but you and a chimpanzee do not. Even if we do not give a set of necessary and sufficient conditions for two things to have the same basic genotype, we can still employ a more casuistical approach, by giving paradigm cases and by using analogical reasoning.[45] That is, we can give paradigm cases of two things having the same basic genotype, paradigm cases of two things *not* having the same basic genotype, and comparison cases that are more similar to the latter than the former. I tend to think that teratomas and hydatidiform moles are themselves paradigm cases of the latter sort. But even if you dispute this, I think that teratomas and hydatidiform moles are more similar

---

[44] But especially Christopher Tollefsen, in personal correspondence.

[45] For an overview of one way of incorporating casuistical approaches into an overall methodology in bioethics, see Beauchamp and Childress (2008, Chapter 10).

to the paradigm cases of the later sort than they are to the paradigm cases of the former sort.

Second, a similar approach might be useful in considering the products of ANT-OAR. The process of "Altered Nuclear Transfer" (ANT) was described to the President's Council for Bioethics by William C. Hurlbut on December 3, 2004, and was discussed in that Council's eventual White Paper Report in May 2005 titled "Alternative Sources of Human Pluripotent Stem Cells." As they summarized the procedure before considering objections:

> ANT, the modified procedure proposed by Hurlbut, involves altering the somatic cell nucleus *before* its transfer to the oocyte, and in such a way that the resulting biological entity, while being a source of pluripotent stem cells, *would lack the essential attributes and capacities of a human embryo*. For example, the altered nucleus might be engineered to lack a gene or genes that are crucial for the cell-to-cell signaling and integrated organization essential for (normal) embryogenesis. It would therefore lack organized development from the very earliest stages of cell differentiation. Such an entity would be a "biological artifact," not an organism.[46]

As worded, this explanation seems to *automatically* rule out the idea that products of ANT would have the capacities that generate serious moral status, since these products "lack the essential. . .capacities of a human embryo." In addition, however, a kind of case-based reasoning is also available:

> In offering his proposal for ANT, Hurlbut emphasizes that no embryo would ever be created or destroyed; since the genetic alteration is carried out in the somatic cell nucleus before transfer, the biological artifact is "*brought into existence* with a genetic structure insufficient to generate a human embryo." Hurlbut compares the product of ANT to certain ovarian teratomas and hydatidiform moles, genetically or epigenetically abnormal natural products of failed fertilization that are not living beings but "chaotic, disorganized, and nonfunctional masses."[47]

A third point is important here, although it tends to be neglected in such discussions. The claim that there *is* a line between "defective" human beings, and entities so deficient that they should not be considered to be human beings, does *not* depend on being able to say *where* that line is. Similarly, the claim that there *is* such a line does not even depend on us being able to say *how* we might find out *where* that line is. Consider a parallel: we may be completely clueless about the location of something, and even completely clueless as to how to find the location, and yet this double cluelessness on our part does not affect in the least whether the thing has a location and what that thing's location is.

However, this brings us to a fourth point, which is even more neglected in such discussions than the previous point. Even if we are not currently able to say *where* the aforementioned line is between defective humans and nonhuman entities, I believe we do know *how* we might find out where that line is. Our empirical method in this case could parallel our empirical method concerning the process of death. First, consider those conditions which must be met for a human organism to have

---

[46] The President's Council on Bioethics (2005, p. 37).

[47] The President's Council on Bioethics (2005, p. 38).

the sort of Near-Death Experience that we described as an Out-of-Body Experience (OBE). Call these conditions "the disembodiment conditions" for a human at the end of her biological life. Next, consider the conditions that complement the disembodiment conditions: that is, consider the conditions which must be met for a human organism *not* to have an OBE. Call those complementary conditions the "embodiment conditions" for a human being at the end of her biological life. Whatever the most general embodiment conditions turn out to be for a human being at the *end* of her biological life, we should assume that the embodiment conditions are the same for a human being at the *beginning* of her biological life.

For example, imagine we discover that brain process B turns out to be a general embodiment condition for a human being at the end of her biological life. This might be tested empirically by observing patients whose brain process B stops, and then starts up again, with the patient reporting out-of-body experiences during the time interval in between. The upshot is that any analogous type of event at the beginning of life would also be an embodiment condition. Let's say brain process B is an instance of a more general biological process P. Instances of P might be found both in things with brains and in things without brains. Let's say instances of P are found in the products of ANT-OAR but not in teratomas or hydatidiform moles. Then the entities created by ANT-OAR would be examples of human organisms.

A similar methodology can be brought to questions of anencephaly. A representative statement concerning the meaning of this condition is contained in the 1996 document "Moral Principles Concerning Infants with Anencephaly" from the National Conference of Catholic Bishops:

> Anencephaly is a congenital anomaly characterized by failure of development of the cerebral hemispheres and overlying skull and scalp, exposing the brain stem. This condition exists in varying degrees of severity. Most infants who have anencephaly do not survive for more than a few days after birth. Modern medical techniques usually can determine this condition with a high degree of certainty before birth.[48]

As I discussed earlier in my treatment of John Rawls, I think the anencephalic human infant is best viewed as a case of a human organism with the higher-order capacity to think. An objection I constantly meet to my approach is that my way of handling the anencephalic is a significant stretch. "Of course," it is objected, "it may be true that an anencephalic *could* someday think, but only in light of a very broad definition of the word 'could': it would essentially require that we have a way to transplant brains, or help anencephalics grow brains, or rely on some kind of artificial brain. But if you are willing to go this far in defining the word 'could', then some of the limits you want to set up elsewhere seem inappropriate (e.g. corpses and animals)".[49] Let me be clear here: I am willing to go this far in defining the word "could." And the limits I want to set up elsewhere are, I believe, justified by the

[48]Committee on Doctrine, National Conference of Catholic Bishops (1996).

[49]Thanks are due to an anonymous reviewer for being one of my more forceful, clear, and charitable critics on this particular point. I owe the wording in the text largely to this reviwer's formulation of the objection.

methodology of reflective equilibrium, especially when the arguments elsewhere in this book are taken into account.

Still, I recognize that for many people, my treatment of anencephaly is a hard pill to swallow. At least before the publication of the Document quoted a moment ago, there was some intramural debate among Catholics regarding anencephalics. Benedict Ashley, in his "Observations on the Document," poses the skeptical challenge this way (before eventually answering it):

> In the controversy over abortion, questions have often been raised about the "hominization" or "individuation" of the embryo or foetus in its early stages. As already noted, in conformity to current embryology, magisterial documents now affirm that human life begins at conception. But, since there is some evidence that anencephaly in some cases may have a genetic cause or predisposition, may it not be that this genetic defect is so radical that, at the very point of conception, the organism so affected never receives a human soul? Thus the false pregnancy producing a "hydatidiform mole" is known to be the result of a failure of the process of conception itself with the result that the normal sequence of development is never initiated.[50]

Ashley's own answer to this skeptical challenge reinforces the position taken by the Document: anencephaly is *less* like the (paradigm) case of hydatidiform moles, and *more* like the (paradigm) cases of undisputed human organisms:

> In anencephaly, however, normal development takes place up to a critical point of organic differentiation. It is more comparable to such cases as the "thalidomide babies" who because of chemical interference at a critical point in their development failed to develop normal limbs, but many of whom have reached adulthood, and who are certainly persons from conception. Anencephaly belongs to the class of physical defects in which for various causes, including possibly some genetic predisposition, the neural tube fails to close. Some cases, however, are now known to be due to easily preventable vitamin deficiencies. Thus there is no sufficient reason to doubt that the anencephalic infant began life as a person.[51]

One of my reasons for citing these Catholic documents is to emphasize how their methodology in approaching the anencephalic is a methodology that makes room for case-based comparisons and analogies. But another reason is to reiterate my conviction that there is a way of going beyond the range of cases typically considered in such case-based methodologies. Once again, a more empirical approach to the end of life might shed light on the human anencephalic. Imagine once again that we empirically discover a brain process B which turns out to be a general embodiment condition for a human being at the end of her biological life. As before, this might be tested empirically by observing patients whose brain process B stops, and then starts up again, with the patient reporting out-of-body experiences during the time interval in between. And once again, the upshot is that any analogous type of event at the beginning of life would also be an embodiment condition. Let's say brain process B is an instance of a more general biological process P. Instances of P might be found both in things with brains and in things without brains. Let's say instances of P are found in thalidomide babies but not in anencephalics. Then

[50]Benedict Ashley (2009).

[51]Benedict Ashley (2009)

thalidomide babies would be examples of human organisms, whereas anencephalics would not. Conversely, if instances of P are found in both thalidomide babies and anencephalics, then both would be examples of human organisms.

At this point, some readers may be wondering: yes, perhaps these anencephalics are human organisms. But is it not difficult to believe, at the end of the day, that such human organisms have serious moral status? One way of putting this objection runs as follows. "Granted, futuristic hypothetical technology could enable the anencephalic to think. But why should we care so much about futuristic hypothetical technology? After all, it seems that there is an important difference between saying: (1) If all proceeds normally, the 6 week old embryo will develop into a fetus and then into an infant and then eventually be able to think and (2) if all proceeds as expected and we do everything possible to help it, the 6 week old embryo will develop into a fetus with anencephaly and will be born and become an infant who, no matter what we do, never will be able to think. So, assume the best of intentions and efforts, and you still are unable to think. This is different from the kid who will be lynched no matter what. Perhaps once we push the concept of "could" too far, we will be left saying: Sure, in theory that human being "could" think with proper assistance, but that assistance is not forthcoming and so there is no morally meaningful sense in which the person has the capacity to think. If you want to tell me why I ought not to kill that human being, you better come up with a better reason."[52]

I believe that this objection can be answered. There are several reasons we should care about futuristic hypothetical technology. One is that it forces us to think about which things have which capacities now. If the locus of capacities really is the locus of serious moral status, then thinking about future technologies helps us to find those things with serious moral status now. But a related reason for thinking about future technologies is that our rationale and motivation for developing certain future technologies rather than others is closely connected with our views about serious moral status. Consider a simple example. Are humans in persistent vegetative states best viewed as potential sources of available organs for transplant or as potential recipients of reconstructive brain surgery? The way we answer will have an impact upon which technologies are developed: transplant technology in the former case, reconstructive brain technology in the latter. Even if there is nothing at all we can do at present to restore the brain functions of those in PVS today, the fact that PVS patients still have serious moral status should motivate us to (among other things) continue searching for ways to bring PVS patients back to normal functioning using reconstructive brain surgery and other innovations.

Part of this objection, however, relies on a misunderstanding. The structure of my argument is not: because we can conceivably heal this individual, it follows that they have serious moral status. Rather, my argument has been: because we can conceivably heal this individual, this shows that they still have certain capacities, and it's the fact that they have these capacities that gives them serious moral status. This objection's reference to a "morally meaningful sense in which the person has

---

[52]Thanks to an anonymous reviewer for really pushing me with this objection.

the capacity to think" seems to be relying on an assumption that unless we can do something now to affect whether a given capacity to think gets exercised, then that capacity to think is not morally meaningful. I believe this is a mistaken assumption. The lynching case from Chapter 3 is indeed different in many ways from the case of an anencephalic child living in 2009, but the point made by the lynching case still stands: the serious moral status of an entity is not merely a function of what consequences we can presently bring about for that entity.

At the same time, nothing that has been said in this book should be construed as insisting that an entity's serious moral status always implies that the entity must be kept alive. I have studiously tried to avoid saying too much about the concept of serious moral status. As I said in the opening paragraphs of the Introduction, and again in Section 3 of Chapter 1, serious moral status is a place-holder for whatever morally salient features normal adult human persons, like you and I, possess. I left it open exactly what these features were. Just because something has serious moral status, this does not automatically mean that heroic life-saving measures must be taken to prolong its life.

In this connection, let me close this book with the words of Benedict Ashley, the Catholic moral theologian whose views on anencephaly were quoted a few pages ago. Although I am not a Roman Catholic myself, I believe that Ashley's words eloquently express my view that the child with anencephaly, like the rest of us, has serious moral status. And yet they also convey my view that the admission that a person has serious moral status does not automatically settle many other important questions about exactly what our duties are towards that person:

> Yet the conclusion that the child with anencephaly has a right to ordinary care only and his or her caretakers only an obligation to give the child that kind of care, does not answer the question, "What is obligatory ordinary care?". This question has been much debated among Catholic moralists, but must be answered in the light of the principle enunciated in the *Declaration on Euthanasia* and confirmed in The Gospel of Life that obligatory treatment and care must be judged by the proportion between the benefit to the child and the burden placed on the care-givers. Obviously this principle requires a prudential judgement, and it seems excessively casuistic to try to reduce it to further absolute norms, since the crucial question is "benefit to the person" and "benefit" is not easy to define. Some argue that since human life is so great a good, any care that prolongs life is an infinite benefit. The Catholic moral tradition has, however, always considered physical life, although good in itself, not as absolute but as a good subordinate to the spiritual good of the person. The conclusion is that if any procedure of care or medical treatment cannot enable a person, including the child with anencephaly, to have at least the possibility of performing spiritual acts it is not of significant benefit and cannot be morally obligatory. This conclusion, of course, in no way justifies the deliberate shortening or termination of human life, but only relieves the caretakers from the obligation of forms of treatment or care which they prudently judge to be of no significant benefit to the patient, even if these are urged by medical professionals or others. The child with anencephaly, therefore, must be so cared for that he or she is as free of suffering as possible and should be given loving attention, affection and protection so that it will be evident to all that she or he is a child like other children.[53]

---

[53]Benedict Ashley (2009)

Bibliography

Aristotle *Metaphysics*. Translated by W. D. Ross. In *The Basic Works of Aristotle*, ed. R. McKeon. New York: Random House (1941).

Armstrong, D. M., C. B. Martin, and U. T. Place. *Dispositions: A Debate*, ed. T. Crane. New York: Routledge (1996).

Ashley Fr. Benedict, O. P., "Moral Principles Concerning Infants with Anencephaly: Observations on the Document." Accessed online, May 2009, at http://www.ewtn.com/library/PROLIFE/bcdanen3.htm

Ayer, A. J. "My Death," reprinted in John Donnelly, *Language, Metaphysics, and Death* (2nd edition). New York: Fordham University Press, pp. 226–236 (1994).

Ayer, A. J. "'What I Saw When I Was Dead': Intimations of Immortality." *National Review* (14 Oct. 1988).

Beauchamp, T. and J. F. Childress. *Principles of Biomedical Ethics* (6th edition). Oxford: Oxford University Press (2008).

Becker, L. "The Priority of Human Interests." In *Ethics and Animals*, eds. H. Miller and W. Williams. Clifton, NJ: Humana (1983).

Benn, S. "Abortion, Infanticide, and Respect for Persons." In *The Problem of Abortion* (2nd edition), ed. J. Feinberg. Belmont, CA: Wadsworth, pp. 135–144 (1984).

Boonin, D. *A Defense of Abortion*. Cambridge: Cambridge University Press (2003).

Broad, C. D. *Examination of McTaggart's Philosophy*, Vol. 1. Cambridge: Cambridge University Press (1933).

Buckle, S. "Arguing from Potential." In *Embryo Experimentation: Ethical, Legal and Social Issues*, eds. P. Singer et al. Cambridge: Cambridge University Press, pp. 90–108 (1990).

Butler, K. "The Moral Status of Smoking." *Social Theory and Practice* 19(1): 1–26 (Spring 1993).

Charo, R. A. "Every Cell is Sacred: Logical Consequences of the Argument from Potential in the Age of Cloning." In *Cloning and the Future of Human Embryo Research*, ed. P. Lauritzen. Oxford: Oxford University Press, pp. 82–92 (2001).

Committee on Doctrine, National Conference of Catholic Bishops 1996. *Moral Principles Concerning Infants with Anencephaly*. September 19, 1996. Accessed online May 2009 at http://www.usccb.org/dpp/anencephaly.htm.

Conn, C. H. "Female Genital Mutilation and the Moral Status of Abortion." *Public Affairs Quarterly* 15(1): 1–15 (January 2001).

Couper, R. "Correspondence Regarding Near-Death Experiences." *The Lancet* 359: 2116 (June 15, 2002).

DeGrazia, D. *Human Identity and Bioethics*. Cambridge: Cambridge University Press (2005).

DiSilvestro, R. "Capacities, Hierarchies, and the Moral Status of Normal Human Infants." *Journal of Value Inquiry* 43: 479–492 (December 2009).

DiSilvestro, R. "A Qualified Endorsement of Embryonic Stem-Cell Research, Based on Two Widely Shared Beliefs about the Brain-Diseased Patients Such Research Might Benefit." *Journal of Medical Ethics* 34: 563–567 (July 2008).

R. DiSilvestro, *Human Capacities and Moral Status*, Philosophy and Medicine 108,
DOI 10.1007/978-90-481-8537-5, © Springer Science+Business Media B.V. 2010

DiSilvestro, R. "Not Every Cell is Sacred: A Reply to Charo." *Bioethics* 20(3): 146–157 (June 2006).

DiSilvestro, R. "Human Embryos in the Original Position?" *Journal of Medicine and Philosophy* 30(3): 285–304 (June 2005).

Dombrowski, D. *Babies and Beasts: The Argument from Marginal Cases*. Chicago: University of Illinois Press (1997).

Ereshevsky, M., ed. *The Units of Evolution: Essays on the Nature of Species*. Cambridge, MA: The MIT Press (1992).

Feinberg, J. "Abortion." In *Matters of Life and Death*, ed. T. Regan. Philadelphia: Temple University Press, pp. 183–217 (1980).

Feldman, F. *Confrontations With the Reaper: A Philosophical Study of the Nature and Value of Death*. Oxford: Oxford University Press (1992).

George, R. P. and C. Tollefsen. *Embryo: A Defense of Human Life*. New York: Random House Publishers (Doubleday) (2008).

Gomez-Lobo, A. "Sortals and Human Beginnings." Online paper accessed May 2009 at http://ontology.buffalo.edu/medicine_and_metaphysics/Gomez-Lobo.doc.

Harman, E. "The Potentiality Problem." *Philosophical Studies* 114: 173–198 (2003).

Harman, E. "Moral Status." Ph.D. diss., MIT. Available online at http://hdl.handle.net/1721.1/17645 (2003).

Hoffman, J. and G. S. Rosenkrantz. *Substance: Its Nature and Existence*. London: Routledge (1997).

Harre, R. and E. H. Madden. *Causal Powers*. Oxford: Blackwell (1975).

Kershnar, S. "The Moral Status of Harmless Adult-Child Sex." *Public Affairs Quarterly* 15(2): 111–132 (April 2001).

Kuhse, H. and P. Singer *Individuals, Humans, Persons: Questions of Life and Death*. Herstellung, Germany: Sankt Augustin, Academia Verlag (1994).

Lee, P. and R. P. George "The Wrong of Abortion." In *Contemporary Debates in Applied Ethics*, eds. A. Cohen and C. H. Wellman. Oxford: Blackwell (2004).

Lockwood, M. "Warnock Versus Powell (And Harradine): When Does Potentiality Count?" *Bioethics* 2(3): 187–213 (1998).

McInerney, P. K. "Does a Fetus Already Have a Future-Like-Ours?" *Journal of Philosophy* 5: 267–268 (1990).

McMahan, J. *The Ethics of Killing: Problems at the Margins of Life*. Oxford: Oxford University Press (2002).

Molnar, G. *Powers: A Study in Metaphysics*. Oxford: Oxford University Press (2003).

Moreland, J. P. and S. B. Rae. *Body and Soul: Human Nature and the Crisis in Ethics*. Downers Grove, IL: InterVarsity Press (2000).

Mumford, S. *Dispositions*. Oxford: Oxford University Press (1998).

Nagel, T. *Mortal Questions*. Cambridge: Cambridge University Press (1979).

Narveson, J. "Animal Rights." *Canadian Journal of Philosophy* 7: 161–178 (1997).

Neuhaus, R. J. *The Eternal Pity*. Notre Dame, IN: University of Notre Dame Press (2000).

Neuhaus, R. J. *The Naked Public Square: Religion and Democracy in America*. Grand Rapids, MI: William B. Eerdmans Publishers (1984).

Norcross, A. "Killing, Abortion, and Contraception: A Reply to Marquis." *The Journal of Philosophy* 87(5): 268–277 (1990).

Nussbaum, M. "Capabilities and Human Rights." *Fordham Law Review* 66: 273–300 (1997). Reprinted in pp. 117–149 of eds. P. De Greiff and C. Cronin, *Global Justice and Transnational Politics: Essays on the Moral and Political Challenges of Globalization*. Cambridge, MA: MIT Press (2002).

Nussbaum, M. "Human Functioning and Social Justice: In Defense of Aristotelian Essentialism." *Political Theory* 20: 202–246 (1992).

Nussbaum, M. "Nature, Function, and Capability: Aristotle on Political Distribution." In *Oxford Studies in Ancient Philosophy*, Supplementary Volume I. Oxford: Oxford University Press, pp. 145–184 (1988).

Nussbaum, M. *Women and Human Development*. Cambridge: Cambridge University Press (2000).
Nussbaum, M. *Frontiers of Justice: Disability, Nationality, Species Membership*. Cambridge, MA: The Belknap Press of Harvard University Press (2006).
Nussbaum, M. "Human Dignity and Political Entitlements." In The President's Council on Bioethics, Human Dignity and Bioethics. Washington, DC. The President's Council on Bioethics, pp. 351–380 (March 2008).
Oderberg, D. "Modal Properties, Moral Status, and Identity." *Philosophy and Public Affairs* 26(3): 259–298 (1997).
Olson, E. *The Human Animal: Personal Identity Without Psychology*. Oxford: Oxford University Press (1997).
Parfit, D. *Reasons and Persons*. Oxford: Clarendon Press (1984).
Parnia, S. and P. Fenwick "Near Death Experiences in Cardiac Arrest: Visions of a Dying Brain or Visions of a New Science of Consciousness?" *Resuscitation* 52: 5–11 (2002).
Peters, T. "Embryonic Stem Cells and the Theology of Dignity." In *The Human Embryonic Stem Cell Debate: Science, Ethics, and Public Policy*, eds. S. Holland, K. Lebacqz and L. Zoloth. Cambridge, MA: The MIT Press (2001).
Pojman, L. "The Moral Status of Affirmative Action." *Public Affairs Quarterly* 6(2): 181–206 (April 1992).
The President's Council on Bioethics *Alternative Sources of Human Pluripotent Stem Cells*. Washington, DC: The President's Council on Bioethics (May 2005).
The President's Council on Bioethics *Controversies in the Determination of Death: A White Paper by the President's Council on Bioethics*. Washington, DC: The President's Council on Bioethics (January 2009).
Prior, E. *Dispositions*. Aberdeen: Aberdeen University Press (1985).
Quinn, W. "Abortion: Identity and Loss." *Philosophy and Public Affairs* 13(1): 24–54 (Winter 1984).
Rawls, J. *A Theory of Justice*. Cambridge: Harvard University Press (1971).
Rawls, J. *Political Liberalism*. New York: Columbia University Press (1993).
Rea, M., ed. *Material Constitution: A Reader*. Lanham, MD: Rowman & Littlefield (1997).
Reagan, M. "I'm With My Dad on Stem Cell Research." Accessed online (8/13/2007) at http://www.humaneventsonline.com/article.php?id=4286
Reagan, R., Jr. Speech to the 2004 democratic national convention, accessed online (8/13/2007) at http://www.pbs.org/newshour/vote2004/demconvention/speeches/reagan.htm
Reichlin, M. "The Argument from Potential: A Reappraisal." *Bioethics* 11(1): 1–23 (January 1997).
Regan, T. "An Examination and Defense of One Argument Concerning Animal Rights." *Inquiry* 22 (1979).
Sen, A. "Capability and Well-Being." In *Quality of Life*, eds. Martha N. and Amartya S. Oxford: Clarendon Press, pp. 30–53 (1993).
Sober, E. *Philosophy of Biology*. Boulder, CO: Westview (1993).
Singer, P. *Practical Ethics* (2nd edition). Cambridge: Cambridge University Press (1993).
Stone, J. "Why Potentiality Matters." *Canadian Journal of Philosophy* 17(4): 815–830 (1987).
Tooley, M. *Abortion and Infanticide*. Oxford: Clarendon Press (1983).
Tuomela, R., ed. *Dispositions*. Dordrecht: D. Reidel Publishing Company (1978).
Van Lommel, P., R. van Wees, V. Meyers, and I. Elfferich "Near-Death Experience in Survivors of Cardiac Arrest: A Prospective Study in The Netherlands." *The Lancet* 358 (December 15, 2001).
Virginia Tech Department of Philosophy. Conference advertisement for "The Metaethics of Moral Status: Perspectives on the Nature and Source of Human Value." April 4–6, 2003. Available online at http://www.phil.vt.edu/HTML/events/metaethics2003.pdf.
Warren, M. A. *Moral Status: Obligations to Persons and Other Living Things*. Oxford: Clarendon Press (1997).
Wiggins, D. *Identity and Spatio-Temporal Continuity*. Oxford: Basil Blackwell (1967).
Wiggins, D. *Sameness and Substance*. Cambridge: Harvard University Press (1980).

Wiggins, D. *Sameness and Substance Renewed*. Cambridge: Cambridge University Press (2001).

Wilkerson, T. E. "Natural Kinds." *Philosophy* 63: 29–41 (1988).

Wilkerson, T. E. "Species, Essences, and the Names of Natural Kinds." *Philosophical Quarterly* 43: 1–19 (1993).

Wilson, R., ed. *Species: New Interdisciplinary Essays*. Cambridge, MA: The MIT Press (1999).

Wreen, M. "The Possibility of Potentiality." In *Values and Moral Standing*, ed. T. Attig, D. Callen, and L. W. Sumner. Bowling Green, OH: Bowling Green State University Press, pp. 137–154 (1986).

Index

A

Actual, continuing subjects of experiences
  defined, 79–80
  compared with capacities, 81–85
Adam (teenager who believes he can't survive past nineteen), 30–31
Albert and Benjamin (comparable adult and infant), 43–44, 76, 77
Al (example of re-electing the potential president), 138
Alien abduction example, 84
Altered Nuclear Transfer-Oocyte Assisted Reprogramming (ANT-OAR), 194
Alzheimer's disease, 1
Anencephaly, 196–201
A-rays and B-rays
  that change a human into a dog and a dog into a human, 159–160, 167–168
  that produce and cure a genetic disability, 58–59
Argument from Marginal Cases (AMC), 55–56
  critical *versus* constructive version, 143–144
  weak *versus* strong version, 144
Argument of book, xii
Argument from Potential (AFP)
  controversial, 109
  different versions, 109
  its importance, 109
Aristotle, 18–19
Armstrong, D. M., 17
Ashley, B., 197, 199
Ayer, A. J., 176–178

B

Balaam's donkey, 161–162
Becker, L., 56
Benn, S., 132, 134
Ben (teenager who believes he can't survive his twenties), 31–32
Bill (example of being a person/then not being a person/then being a person again), 138
Boonin, D., 78–79, 147–148, 153
Brain damage and genetic conditions, 57–60
Broad, C. D., 22–23, 56–57
Buckle, S., 121
Bush, G. W., 81–82
Butler, K., 11

C

Caleb and Drake (comparable adult and fetus), 45, 77
Capacities, 184–193
  active *versus* passive, 18–19
  characterized, 17–18
  controversial, xiii
  first-order *versus* immediate/immediately exercisable, 25–26, 73
  general *versus* specific, 19–20
  hierarchies of, 20–26
  identity-preserving *versus* compositional, 18–19, 189–190
  typical human capacities, 19, 26
Caterpillars *versus* scatterpillars, 159
Cerebral cortex (not necessary for having higher-order capacity to think), 40
Charo, R. A., 117–131
Clinton, H., 9–10
Comparison argument
  defended, 76–79
  set out, 77
Complete reprogramming of upper brain (Tooley), 81
Conn, C. H., 11

Corpse problem
described, 169–170
"not dead yet" response, 170
Couper, R., 180

D
Death
cardiopulmonary *versus* neurological criteria for, 170–171
discussed by President's Council on Bioethics, 170–173
often assumed to be irreversible, 170
reversible, 173
Developmental distance, 35–37
Dispositions
and background conditions, 60–61
and initiating causes, 61–62
Dombrowki, D., 56, 92, 145
Dumbo (the elephant), 158

E
Ebert and Frank (comparable adult and embryo), 45, 77
Egoistic concern, 4
Embarrassing Entailment, The, 138
Embodiment conditions *versus* disembodiment conditions, 195–196
Ereshevsky, M., 154
Essential properties
individually-essential, 159
kind-essential, 159

F
Feinberg, J., 27, 133, 134
Feldman, F., 13
Finnis, J., 184
Fission, 52–55
Flatliners (movie), 183
Fusion, 52–55

G
Gandhi, 83
George, R. P., 184–188
Gilbert and Harry (comparable adult and zygote), 45, 77
Gomez-Lobo, A., 192–193
Grisez, G., 184

H
Harman, E., 12, 109
Harre, R., 17
Hoffman, J., 122–123
Human
count sense *versus* stuff (mass, quantity) sense, 122–123
definition (DiSilvestro), 9–10
descriptive concept of, 101
Nussbaum's concept of, 101
various senses of, 8
Hume, D. *versus* Robinson, D., 24
Hurlbut, W., 184, 195
Hydatidiform moles, 194

I
Identical twins (in thought-experiment), 68–69
Incapacities, 158
Innocuous Implication, The, 138
Irreversible damage, 58

J
John (Tooley's temporarily desireless adult), 48

K
Kant, I., 8
Kerry, J., 81–82
Kershnar, S., 11
Killing
and skilling, 15–16
strong moral presumption against, 12–13
Kuhse, H., 52–53

L
Lee, P., 185, 186
Lew and Kareem, 111–112
Lockwood, M., 55

M
Madden, E. H., 17
McInerny, P. K., 117
McMahan, J., 13–14
fetus with cerebral deficits compared to dogs, 148–153
and shrinking to become an embryo, 51–52, 71, 145
Mereological sum, 113–116
Mereological switch, 111–112
Modal boundary, 159
Molnar, G., 17, 25
Moral status
different concepts of, 10–12
independent of historical properties, 70
serious, 12–15
Warren's definition, 5
Moreland, J. P., 123
Mumford, S., 17

N
Nagel, T.
brain bisection, 54
infantilized adult, 43, 57
Narveson, J., 143
Natural kinds, 155–157
and evolution, 161
Near Death Experiences (NDEs), 174
appropriately named, 175
relevance to temporary change arguments, 180–182
Neuhaus, R. J., 175–176
Norcross, A., 115–116
Normalizing approach, 37–38
Nussbaum, M., 65, 85, 94–104
basic capabilities, 95–96
capabilities and human rights, 100–101
capabilities and justice in distribution, 97–100
combined capabilities, 97
differences between, and DiSilvestro's approach, 101–103
internal capabilities, 97
lower-level capabilities the same as higher-order capacities, 96
summary of the relevance of, capabilities approach, 101

O
Obama, B., 9
Oderberg, D., 131
Olson, E., 6, 22
Out of Body Experiences (OBEs), 179–180
Overdetermination, 71, 72–73

P
Parfit, D.
fission of the self, 54
replication booths, 70
Parkinson's disease, 2
Parnia, S. and Fenwick, P., 178–179
Parthenogenesis, 114–115
Passive capacities, 60
Passive potentiality, 46
Pellegrino, E., 182–183
Permanent vegetative state (PVS) applied to genes, 118–119, 126–127
Personal identity
with an anesthetized/sleeping/drunk person, 43
with an infant, 42–43
causal connections not necessary for, 47–51
psychological accounts of, 44
Personal Pronoun Police, 29–30
Personhood* Generates A Right To Life, 141
Personhood
normative/moral personhood *versus* descriptive/commonsense, 27, 133
in potential, 132
Persons, 26–29
Peters, T., 117
Pojman, L., 11
Potential to become *versus* potential to produce, 121
Potential Personhood Generates A Right To Life, 132
Potential persons
Tooley's account, 45–46
why it's better to avoid the concept, 137
Potential president example and the problem of re-election, 131–241
Potential Relevant To Moral Status, 132
President's Council on Bioethics
discussing alternative sources of human pluripotent stem cells, 195
discussing death, 170–173
Prior, E., 17, 60–64
Proteus, 159
Python, M., 118

Q
Quinn, W., 130

R
Rae, S., 123
Rawls, J., 85–94
higher-order capacities as the basis of moral personhood and human equality, 89–94
higher-order capacities in *Political Liberalism*, 87–88
paternalism and the original position, 86–89
Reagan, M., 1–2
Reagan, R., 1–7, 32–33, 47–48, 167
Reagan, R. Jr., 1–2
Rea, M., 112
Regan, T., 143–144
Reichlin, M., 117, 119
Replicas, 70–71, 72–73
Rosenkrantz, G., 122–123

S
Sabom, M., 180
Scatterpillars, 159–160
Scrooge-like objection, 60–64
Sen, A., 94

Serious moral status, definition of, xi, 10–15
Shrinking human example (McMahan), 51–52, 114, 116–117
Shumans and humans, 66–67, 164
Singer, P., 52–53, 132
Sober, E., 155, 161
Sortals, 16, 18, 135–136, 192
  "person" as phase or substance sortal, 136–137
  phase sortals *versus* substance sortals, 18, 136
  "president" as phase sortal, 135–136
Species problem, 154–155
Stone, J., 109

T
Temporary change argument
  metaphysical version, 39–42, 165–166
  moral version, 66, 68, 166–167
Temporary changes
  assumptions, 39
  defined, 39
Temporary coma
  applied to genes, 118–119, 126–127
Teratomas, 194
Tollefsen, C., 184–188
Tooley, M., 109–110, 129
  distinctions among capacities/potentialities, 20–22
  approach to temporary changes involving "deprogramming" of the upper brain, 46–47
  kitten example, 147, 153
  and persons, 28–29, 45
  reasons for not trusting common intuitions about infanticide, 66
Totipotency, 52–55
Tuomela, R., 17

U
Van Lommel, P., 179–180

V
Warnock, M. A., 55
Warren, M. A., 5, 92, 113–114
Wiggins, D., 18, 136
Wilkerson, T. E., 155–157
Wilson, R., 154
Wreen, M., 109, 135

Z
Zygotes, 42, 45, 53, 78, 113*ff*

LaVergne, TN USA
28 June 2010
187594LV00002B/5/P

9 789048 185368